Equations from God

Johns Hopkins Studies in the History of Mathematics
Ronald Calinger, Series Editor

Equations from God
Pure Mathematics and Victorian Faith

DANIEL J. COHEN

The Johns Hopkins University Press
BALTIMORE

© 2007 The Johns Hopkins University Press
All rights reserved. Published 2007
Printed in the United States of America on acid-free paper
2 4 6 8 9 7 5 3 1

The Johns Hopkins University Press
2715 North Charles Street
Baltimore, Maryland 21218-4363
www.press.jhu.edu

Library of Congress Cataloging-in-Publication Data
Cohen, Daniel J.
Equations from God : pure mathematics and victorian faith / Daniel J. Cohen.
p. cm. — (Johns Hopkins studies in the history of mathematics)
Includes bibliographical references and index.
ISBN-13: 978-0-8018-8553-2 (hardcover : alk. paper)
ISBN-10: 0-8018-8553-1 (hardcover : alk. paper)
1. Religion and science. 2. Mathematics. I. Title.
BL265.M3K64 2007
261.5′5—dc22   2006019751

A catalog record for this book is available from the British Library.

For Rachel ∞

Contents

Acknowledgments

A welcome side effect of pursuing the history of ideas is that it makes you acutely aware of your own intellectual indebtedness. I have been blessed with a string of outstanding advisors and mentors: John Williams and Donald Burke at Weston High School in Weston, Massachusetts; Gerald Geison at Princeton; David Hall at Harvard; and especially Frank Turner at Yale. Frank saw this book through from its inception to completion with unparalleled insight into the Victorian era, helpful criticism, and warm congeniality. Jon Butler's infectious enthusiasm, shrewd guidance, and astute sense of religious history were also enormously important. I profited greatly from the comments of others at Yale, including David Brion Davis, Harry S. Stout, and Louise Stevenson.

I completed this book at George Mason University, which has managed to assemble a history department filled with imaginative scholars who also happen to be uniformly nice. Jack Censer, former department chair and now dean, undoubtedly deserves a great deal of credit for this remarkable environment; I appreciate his friendship as well. Daniele Struppa, the former dean and a mathematician, deserves thanks for championing my work on this and other projects.

Every day I enjoy the intellectual stimulation and camaraderie of my colleagues at the Center for History and New Media, directed by Roy Rosenzweig. I greatly treasure Roy's friendship and advice. At the Center, Olivia Ryan and Josh Greenberg kindly helped with final edits to the manuscript.

I also deeply appreciate the encouragement and assistance of Trevor Lipscombe, Henry Y. K. Tom, and the staff at the Johns Hopkins University Press, especially Nancy Wachter and Erin Cosyn. They have been a pleasure to work with. I am grateful to Jim O'Brien for his preparation of the index.

This book would not exist without generous financial support from a variety of sources. In its research stage, a John F. Enders grant, a Mellon research grant, and a fellowship from The Pew Charitable Trusts' Program in American Religious History enabled me to visit critical archives in the United States and Great Britain. After my research was complete, I had the great luxury of writing an initial version of this study on a Charlotte W. Newcombe Fellowship from the Woodrow Wilson Foundation.

Less tangible, but even more important, has been the lifelong support of my mother, father, and sister. My son and daughter, Arlo and Eve, have provided much needed comic relief.

My greatest inspiration has been my wife, Rachel, to whom I dedicate this book. I see in her what the early Victorians saw in mathematical equations—a combination of wondrous elements from the earthly and heavenly realms.

Equations from God

The Allure of Pure Mathematics in the Victorian Age

On September 23, 1846, the Berlin astronomer Johann Gottfried Galle scanned the night sky with a telescope and found what he was looking for—the faint light of the planet Neptune. Excitement about the discovery of an eighth planet quickly spread across Europe and America, generating a wave of effusive front-page headlines. Within scientific circles, however, the enthusiasm rapidly soured into a dispute over who should receive credit. Prior to Galle's search, a young British mathematician named John Couch Adams and a well-known French mathematician named Urbain Le Verrier each had prognosticated the size and position of the planet. Unsurprisingly, the debate over credit quickly acquired a fierce, nationalistic overtone. Many British astronomers and mathematicians saw the discovery as an important opportunity to achieve recognition of their growing acumen, embodied in Adams. Across the channel, Le Verrier had the gall to suggest that the scientific academies name the planet after him, and his Gallic colleagues discounted the role of Adams's work in the actual search for the planet.[1]

This divisive argument obscured what was perhaps the more significant aspect of the planetary discovery: Neptune was the first heavenly body found by mathematical prediction. Without peering into the sky at all, Adams and Le Verrier independently calculated the location of the planet through geometrical analysis and the laws of gravitation. Beginning with extremely precise observations of Uranus's orbital irregularities, each mathematician generated a formula for the planet's deviations from a proper ellipse. Meshing Newton's laws with this mathematical description of Uranus's course, they extrapolated outward to the assumed eighth planet, solving the combined equations for Neptune's mass, motion, and distance from the Sun. Aside from the initial observations of anomalous gravitational perturbations in the orbit of Uranus, the discovery of Neptune was an exercise in pure thought.

This remarkable aspect of the discovery was not lost upon contemporaries. To many other scientists—and many nonscientists as well—the work of Adams and Le Verrier signaled a new era of human knowledge, and they loudly sang its praises. Robert Harry Inglis, the president of the British Association for the Advancement of Science, told those convening in the theater of Oxford University on June 23, 1847, that the past year "had been distinguished by a discovery the most remarkable, perhaps, ever made as the result of pure intellect exercised *before* observation, and determining *without* observation the existence and force of a planet; which existence and which force were subsequently verified *by* observation."[2] To determine a truth without the use of the common senses was for Inglis and others a mark of greatness.

John Herschel, the president of the Royal Astronomical Society, proclaimed that the discovery "surpassed, by intelligible and legitimate means, the wildest pretensions of *clairvoyance*,"[3] and wrote that "the movement of the planet had been felt (on paper, mind) with a certainty hardly inferior to ocular demonstration."[4] He further emphasized the universal character of mathematics:

> That a truth so remarkable should have been arrived at by methods so different by two geometers, each proceeding in utter ignorance of what the other was doing, is the clearest and most triumphant proof which could have entered into the imagination of man to conceive, of the complete manner in which the Newtonian law of gravitation stands represented in the formulæ of those great mathematicians who have furnished the means by which alone this inquiry could have been entered on; and how perfect a picture—what a daguerreotype—those formulæ exhibit of its effects down to the least minutiæ![5]

For Herschel, in other words, Adams and Le Verrier had acted as two independent eyes that in tandem produced a binocular, three-dimensional vision of the distant body of Neptune. No less significant was the fact that the two hailed from different countries and different cultures—a true sign of the genius of mathematics. This transnational characterization of the method behind the planetary discovery thus ran counter to the nationalist debate over proper credit. In France, the physicist and mathematician Jean-

Baptiste Biot echoed Herschel's appeal to the universal aspect of such mathematical analysis: "Minds dedicated to the pursuit of science belong, in my eyes, to a common intellectual nation."[6] Transcendental, unifying truth, Herschel and Biot believed, is available to great minds everywhere through the use of mathematics, which disregards all human boundaries.

Some descriptions of the event went even further, characterizing Adams and Le Verrier as potent sorcerers who had conjured and commanded the supreme realm of Truth. In a highly dramatic passage recalling the Romantic poetry of Coleridge and Wordsworth, the Scottish optics researcher David Brewster declared the superiority of these mathematicians over mere observers:

> [The mathematician] calculates at noon, when the stars disappear under a meridian sun. He computes at midnight, when clouds and darkness shroud the heavens; and from within that cerebral dome which has no opening heavenward, and no instrument but the eye of reason, he sees, in the agencies of an unseen planet, upon a planet by him equally unseen, the existence of the agent; and from the direction and amount of its action he computes its magnitude and place. If man ever sees otherwise than by the eye, it is when the clairvoyance of reason, piercing through screens of epidermis and walls of bone, grasps, amid the abstractions of number and quantity, those sublime realities which have eluded the keenest touch, and evaded the sharpest eye.[7]

At work the mathematician becomes a pure spirit, rising out of the confinement of his material body, Brewster conceived. He has no use for everyday faculties like sight, but rather operates with a higher, far more powerful internal sense. This mathematical faculty is not a passive receptor of information, but is instead a penetrating instrument that attains the grandest truths, all of which lie beyond the reach of our five bodily senses.

In the first published book on Neptune, J. P. Nichol, a professor of astronomy at the University of Glasgow, portrayed the discovery in a similarly dramatic way while portraying the mind of the mathematician as in touch with the underlying properties of the universe. Withholding no superlatives, Nichol (although he was not there) recorded the triumphant moment when Le Verrier announced his calculated position of Neptune to the French Academy on August 31, 1846:

How singular that scene in the Academy! A young man, not yet at life's prime, speaking unfalteringly of the necessities of the most august Forms of Creation—passing onwards where Eye never was, and placing his finger on that precise point of Space in which a grand Orb lay concealed; having been led to its lurking-place by his appreciation of those vast harmonies, which stamp the Universe with a consummate perfection! Never was there accomplished a nobler work, and never a work more nobly done! . . . He trod those dark spaces as Columbus bore himself amid the waste Ocean; even when there was no speck or shadow of aught substantial around the wide Horizon—holding by his conviction in those grand verities, which are not the less real because above sense, and pushing onwards towards his New World![8]

In this histrionic passage, Nichol (like Brewster) demoted the common senses in favor of the higher faculty of mental analysis, which reaches its highest form in pure mathematics. Most mathematicians may seem taciturn, he thought, yet inside they are the true adventurers of the modern age, setting sail for the distant regions of the mind and the universe with powerful ships built from transcendental elements.

The nature of Neptune's unveiling likewise exhilarated William Rowan Hamilton, a prominent Irish mathematician. Indeed, as with Brewster and Nichol, the discovery of the eighth planet so moved him that standard prose seemed woefully inadequate to describe the moment. Instead, Hamilton wrote a poem to honor Adams and astro-mathematical prediction in general. Deeming the weight of Roman mythology to be appropriate to the gravity of his subject, Hamilton compared the ascension of the mathematician to the liberation of the goddess of wisdom:

> When Vulcan cleft the labouring brain of Jove
> With his keen axe, and set Minerva free,
> The unimprisoned Maid, exultingly,
> Bounded aloft, and to the Heaven above
> Turned her clear eyes.[9]

Furthermore, Hamilton emphasized the beneficence of transcendent mathematical discoveries, which come from gifted intellects yet ultimately are not theirs alone. "Having discovered the new planet as a *Truth*," Hamilton wrote to a friend, "[Adams] has so gracefully disclaimed it as a *Possession*."[10]

Hamilton's colleague Augustus De Morgan similarly admonished his own corner in the nationalist fight between supporters of Le Verrier and supporters of Adams. "We may wish that the complete honour of this great fact had fallen upon the English philosopher," De Morgan noted, "but far beyond any such merely national feeling is our desire that philosophers should recognise no such distinction among themselves. The petty jealousies of earth are things too poor and mean to carry up amongst the stars."[11] The ecumenical nature of mathematics naturally led to a certain humility, for mathematicians understood that they were unveilers, not creators, of important facts.

Across the ocean, American intellectuals were not far behind in their proclamations about the significance of the planetary discovery. Cyrus Augustus Bartol, the Unitarian pastor of West Church in Boston, echoed the sentiments of his British colleagues in an 1847 article intended for a broad audience fascinated by Neptune:

> [The mathematician] scans these perturbed inclinations more exactly, measures their amount, ascends to their adequate cause, and though that cause still lay darkly ranging on, with to earthly vision undiscernible lustre, he yet predicts its place, and course, and time of arrival into the focus of human sight. His prediction is recorded, to be entertained by some, or incredulously smiled at by others. But lo! in due time the stranger comes as announced, to fulfil this "sure prophetic word" of the divinely inspired understanding of man.[12]

With breathless accolades such as this, Bartol framed the discovery in a decidedly spiritual way. Thus the planetary announcement of 1846 was for men like Herschel, Brewster, Nichol, Hamilton, and Bartol indicative of a new level of human understanding and a new world in which mathematical geniuses such as Adams and Le Verrier were the great communicators of truth.

The irony of this story is that while the prognostication of Neptune may have been an exercise in pure thought, this thought was not entirely rational. Both Adams and Le Verrier used the scientific laws of Newton and geometrical analysis, but they also relied on one completely unscientific theory called Bode's Law.[13] In a manner that recalls Greco-Roman harmonic visions of the universe, this law stated that the distances of the planets from the Sun correspond to the series $4 + 3(2^n)$, if we define the Earth as being

10 units away and use an adjusted first term for Mercury. The heavenly bodies in our solar system should thus be found at the intervals 4, 7, 10, 16, 28, 52, 100, 196, 388, and so on. Why incorporate into one's rigorous calculations this unjustified, ungainly rule, the mention of which is a highly effective way to produce unbridled laughter among twenty-first century astronomers? The two nineteenth-century mathematicians faced the daunting problem that the perturbations in the orbit of Uranus could be caused equally by a small body close to the planet or a large body distant from it, or any size and distance combination in between. Where was Neptune in this enormous spectrum of physical possibility? To help find the needle in the haystack, both Adams and Le Verrier consulted Bode's Law, which was well respected at the time though difficult to square with scientific method. (It is true, however, that the series produced by the law is uncannily accurate for the first seven planets if one includes the asteroid belt as the fifth term.) The rule forecast a planet at a distance of 388 units from the Sun (38.8 astronomical units, or AU, in modern astrophysical terminology), and this number aided the mathematicians in calculating the size (and thus brightness) of Neptune and its position in the sky. As it turns out, Neptune is actually only 30 AU from the Sun (almost a billion miles closer than believed in 1846), and its mass is just one-half of that predicted by Adams and Le Verrier.

Social perception is often more important than reality, however. Many intellectuals saw the discovery of Neptune as a complete and total triumph of the pure thought of mathematics, and this was its true legacy for the next quarter of a century. The eminent American mathematician Benjamin Peirce spoke for many of his colleagues when he reveled in "facts of which the knowledge is wholly mental, and of which there is no direct evidence to the senses," and he saw these facts as "directly known only to the few who have the logical training to follow the argument by which they are demonstrated; and indirectly to those other few who have the loyal faith to trust the testimony of the geometers."[14] The story of Neptune's unveiling illustrates well the manifestation of such idealism in the work and thought of early Victorian mathematicians and scientists. The discovery of Neptune was for J. P. Nichol and his contemporaries an "ever-memorable adventure into that region of pure thought," a transcendent journey into the land of the fundamental ideas of our universe.[15] Praising the ideal nature of the language of mathematics, Nichol highlighted the fact that Le Verrier and Adams had used "the symbols and processes of our most recondite Analysis," which alone can access invisible, eternal laws.[16]

Such sentiments were obviously more than paeans to mathematics; they were strong professions of a peculiar kind of Victorian faith. As the British astronomer and mathematician Mary Somerville recalled in her autobiography, "Nothing has afforded me so convincing a proof of the unity of the Deity as these purely mental conceptions of numerical and mathematical science which have been by slow degrees vouchsafed to man . . . all of which must have existed in that sublimely omniscient Mind from eternity."[17] Somerville thought herself extremely lucky to have had a career that dealt daily with the divine forms of pure mathematics. A contemporary of Somerville, the Royal Society fellow Henry Christmas, even argued that the study of mathematics was essential to a complete and rich spiritual life. "He who undertakes to be a missionary of Divine truth, must be a man of enlarged and cultivated faculties," Christmas wrote in his book *Echoes of the Universe: From the World of Matter and the World of Spirit* (1850), "Now there is one class of study which we wish to recommend as very important . . . and this is the study of mathematics."[18] For Christmas, the study of mathematics was part of "sanctified learning," an "intellectual blessing" from God not to be overlooked.[19]

With such intense religious meaning attached to the mathematical prognostication of Neptune, the discipline of mathematics quickly became fodder for sermons. In an 1848 oration, the American Congregational minister Horace Bushnell declared that mathematics clearly consisted of "those pure and incorruptible formulas which already were before the world was, that will be after it, governing throughout all time and space, being, as it were, as integral part of God."[20] The symbols and correspondences of mathematics thus "put the mathematician in profound communion with the Divine Thought."[21] Although he was not a mathematician, the religious idealism of his scientific brethren was encouraging to Bushnell. Revelations from the divine sphere comprise the epiphanic moments of science, he believed, as mankind communes with God's great mathematical laws and concepts. "Geometrical and mathematical truths become the prime sources of scientific inspiration; for these are the pure intellectualities of all created being," Bushnell proclaimed.[22] At times of discovery the scientist is "raised to a pitch of insight and becomes a seer, entering into things through God's constitutive ideas, to read them as from God."[23] Without comprehending the equations of Adams and Le Verrier, Horace Bushnell nevertheless could understand and relay how their work invoked the heavenly realm.

In an 1850s sermon, the Oxford clergyman Adam S. Farrar also diverged from his normal subject matter to inform his audience of the profound significance of pure mathematics. "If any branch of knowledge appeared eminently unlikely to unfold to us any information about God, you would think it would be that system of symbolic formulæ and abstract notions," he noted, "And yet when we apply it to predict the attractions of the heavenly bodies in periods yet to come, it unfolds to us some results of extraordinary grandeur."[24] Farrar therefore concluded that the equations of mathematics ultimately "reveal to us the infinite wisdom of God."[25] "Who can contemplate these amazing results, which manifest the infinite contrivance of the Almighty Architect, without a feeling of devout thankfulness that we have been permitted thus to discover traces of the high and lofty One who inhabiteth eternity!" he declared.[26]

Edward Everett, the New England politician, Harvard administrator, and orator, summarized the feelings of many early Victorian clergymen and mathematicians alike in an 1857 lecture at the inauguration of Washington University in St. Louis. He eloquently announced to the spectators, "In the pure mathematics we contemplate absolute truths, which existed in the Divine Mind before the morning stars sang together, and which will continue to exist there, when the last of their radiant host shall have fallen from heaven."[27] Much of Everett's audience surely nodded in agreement with his lofty assessment of mathematics.

This commingling of the mathematical with the spiritual was not exactly new. Western thought had long given the discipline a lofty spot in the pantheon of knowledge. Indeed, since the height of ancient Greece philosophers have often considered mathematics so sublime that it transcends the profane realm of humanity and ascends into the pure realm of the divine. Chapter 1 traces this link between religion and mathematics in the Western intellectual tradition, concentrating on those thinkers who most frequently appeared in the writings of Victorian mathematicians. Plato's assessment of mathematics, particularly in his later works, created a transcendental aura around the discipline, and Platonists from Proclus onward strengthened this sense of the ideal nature of mathematics. In the early modern period, the Cambridge Neoplatonists firmly established this philosophy of mathematics in the English-speaking world. German philosophical idealism flowing from Immanuel Kant and his English disciples, such as William Wordsworth and Samuel Taylor Coleridge, further prompted mathematicians and intellectuals across the Anglo-American world to subscribe to a transcendental philosophy of mathematics.

Perhaps the most robust form of this mathematical idealism flourished in nineteenth-century America. Chapter 2 focuses on Benjamin Peirce (1809–1880), in many ways the founder of pure mathematics in the United States, and his circle. Peirce, the father of the philosopher Charles Sanders Peirce, was a professor at Harvard for almost a half-century, and was close friends with many of the key intellectual figures of Victorian New England. Living in a land with greater religious latitude—and greater religious intensity—than Great Britain, Peirce expanded upon the sentiments aroused by the discovery of Neptune far more deeply and publicly than his British counterparts. His forthright combination of religious idealism with pure mathematics provides the most vivid picture of this link, and exposes many of the features of nineteenth-century faith that made this combination possible and influential.

Despite Peirce's international renown, however, most of the innovative work in pure mathematics continued to come out of the Old World. Existing mathematical fields diversified and new fields arose in response to significant breakthroughs. The eastern European mathematicians János Bolyai and Nikolai Ivanovich Lobachevsky formulated non-Euclidean geometry, a set of principles counterintuitive to normal human experience that led to a complete redefinition of this most ancient of mathematical pursuits. Mathematics and its associated methodologies also expanded into realms of knowledge other than the natural sciences (where they had been especially at home in physics and astronomy), often through pioneering work by theorists who began their careers as mathematicians but who branched out later in life. At the same time that this move outward occurred, there was a move inward in nineteenth-century mathematics. Concerns about the foundations of the discipline—an interest in the fundamental nature of mathematical knowledge and the process whereby mathematicians come to conclusions—occupied a significant portion of the research agenda.

British interest in the formal aspects of mathematics was particularly apparent in the growing interest in mathematical, or symbolic, logic. As George Boole (1815–1864), one of the founders of mathematical logic and the subject of chapter 3, summarized the nature of this critical field of pure mathematics, it was "not of the mathematics of number and quantity alone, but of mathematics in its larger . . . truer sense, as universal reasoning expressed in symbolical forms."[28] While some of the most famous Victorian mathematicians, such as Arthur Cayley and James Joseph Sylvester, studied and contributed to many of the abundant research topics, the British showed an unusually strong interest in this budding field of symbolic logic. A remarkable three generations furthered the association between British thought and logic while creating a new mathematical field in concert with European counterparts: Boole and Augustus De Morgan (1806–1871), William Stanley Jevons (1832–1885) and John Venn (1834–1923), and Bertrand Russell (1872–1970) and Alfred North Whitehead (1861–1947).

The 1840s and 1850s saw the groundbreaking publication of Boole's *The Mathematical Analysis of Logic, Being an Essay Towards a Calculus of Deductive Reasoning* (1847) and *An Investigation of the Laws of Thought on Which Are Founded the Mathematical Theories of Logic and Probabilities* (1854), as well as De Morgan's *Formal Logic* (1847). Jevons, a student of De Morgan at University College, London, began his career by formulating his own symbolic logic (*Pure Logic*, 1864), which led to his landmark treatise *The Principles of Science* (1874), and he continued to work in the field as he carried its methods into economics and the social sciences in general. Venn expanded upon Boole's theories in two critical texts in the 1880s, *Symbolic Logic* (1881) and *The Principles of Empirical Logic* (1889), in the process inventing the diagrams of overlapping shapes that would come to bear his name. *Principia Mathematica* (1910–1913), in which Russell and Whitehead equated logic and mathematics at the deepest level possible, was a culmination of the innovative mathematical research of the Victorian age. Before this seminal collaboration Russell and Whitehead had independently penned monographs exploring mathematical logic (Russell's *The Principles of Mathematics*, 1903; Whitehead's *A Treatise on Universal Algebra*, 1898). Although far from the totality of British mathematics in the nineteenth century, these were among the most highly influential figures in Victorian mathematical circles.

Why did mathematical logic flourish on the British Isles in the second half of the nineteenth century, and why did British mathematicians pursue this particular region of their discipline with such passion? What motivated the founders of mathematical logic, Boole and De Morgan, and why were promising young British mathematicians eager to embrace and extend their work?

These questions lie at the heart of chapters 3 and 4, which investigate the work and faith of Boole and De Morgan. Historians of mathematics and philosophy, who more frequently provide technical accounts of these disciplines rather than branching out into larger contexts, have continually sought to illuminate the noteworthy differences between the systems of these two British mathematicians.[29] Instead I pose the opposite question: What did Boole and De Morgan have in common that would drive them to create this novel technique? It is apparent, in particular, from unpublished sources, that in the middle decades of the nineteenth century Boole and De Morgan were intensely concerned with interfaith agreement in a chaotic era of belief. It is no coincidence that symbolic logic arose in the wake of Catholic Emancipation, the beginning of Jewish emancipation, and the Oxford Movement. In this extratechnical context, the two mathematicians envisioned their logic based on mathematics as a highly ecumenical endeavor. Although philosophers used symbolic logic in the twentieth century as a way to render spiritual questions irrelevant, in the nineteenth century British intellectuals like Boole and De Morgan used it to rise above sectarian boundaries in the name of a true and universal faith. Thus did antidogmatic and antidoctrinal thinkers attempt to provide a basis for a national, even global, union of believers.

This hidden story of the original symbolic logicians undermines the Whig history of Anglo-American philosophy in the century preceding the First World War. That progressive history chronicles the rise of pure mathematics in the early nineteenth century as the source of a reconsideration of the methods of philosophy.[30] Applying new, rigorous, mathematical tools to the subject of logic, a symbolic system arose that held the promise of a clean, certain, scientific philosophy. Toward the end of the nineteenth century mathematical logic slowly diminished interest in traditional modes of philosophy such as idealism, which philosophers had come to see as nebulous and linguistically problematic. Finally, advances based on Russell and Whitehead's *Principia Mathematica* generated a twentieth-century philosophical school founded on dispassionate calculation, grammar, and proofs rather than poorly defined abstract concepts. Philosophy had finally freed itself from theology and confusion.

As we have already seen, it is wrong to assume that the purpose of nineteenth-century pure mathematics and the symbolic logic that arose out of it was to construct a completely scientific, secular realm of philosophy. Boole and De Morgan did not know what the future would hold, and they had very different agendas than the Whig history imagines. A panoramic examination of their writing—not only their mathematical treatises but also their private letters, unpublished works, and even poetry—makes it clear that the creators of symbolic logic and their supporters yearned for a more profound religion than contemporary sects seemed to offer, a religion that did not have its foundation in dogma, liturgy, or ecclesiastical organizations. Mathematical logic would serve God by providing a way to ascend above such human constructions. It is therefore impossible to discuss the work of such mathematicians without referring to the broader history of the middle of the nineteenth century, especially its religious history.[31]

The link between religious idealism and mathematics apparent at mid-century would become problematic, though, as the Victorian age wore on. Mathematical idealism became threatening to many professionals because it broke down important barriers being erected between religion and science, as well as "high" and "low" strata of thought. To a new generation of mathematicians, religiously tinged mathematics seemed uncomfortably close to the efforts of sectarian clergymen, who well after the discovery of Neptune continued to appropriate mathematical notions to advance religious arguments. A divine vision of mathematics also made it harder for professional mathematicians to distance themselves from amateurs who kept pursuing what the professionals saw as anachronistic problems in pure mathematics, such as circle squaring. That contemporary studies of imaginary numbers, four-dimensional algebras, and infinity readily lent themselves to metaphysical or theological interpretations only made this extraction more difficult.[32]

As fatigue over sectarian strife grew in the 1860s and 1870s, and as the discipline advanced with the founding of professional associations and the modernization of curricula, academic mathematicians pragmatically began to isolate themselves from theology. Caught between his yearning for harmonious interfaith relations and the real world of contentious professional concerns, Augustus De Morgan, for one, diverged from the early Victorian understanding of pure mathematics. By fervently advocating religious ecumenism at the same time that he advanced a secularized, humble philosophy of mathematics, De Morgan found a *via media* that permitted his discipline to flourish professionally without disenchanting the universe. By the end of his life De Morgan pronounced that mathematics was not a language from heaven after all; rather it was simply a highly functional, useful science.

De Morgan's inner turmoil over his beliefs and his work anticipated the tensions and contradictions of the late Victorian era, and his solution fore-cast the compromise adopted by the next generation of mathematicians. In-deed, this generation began to *belittle* their own endeavor—by making lesser claims about the nature and parameters of mathematics, they secured the discipline from the prying hands of theologians and amateurs, who were drawn to mathematics because of its supposedly transcendental powers. The consternation interlopers such as amateurs and clerics engendered among these later mathematicians was indicative of a new possessive spirit that obscured the religious sentiment of early Victorian mathematics. Some late Victorian mathematicians even began to rebuke colleagues for includ-ing theological rhetoric in their publications. The grand idealist philosophy of mathematics faded. Rather than claiming that mathematics transcribed the mind of God, late nineteenth-century mathematicians proffered a baser and more pragmatic vision: mathematics was a set of laws and a system of notation created in the *human* mind.

As chapter 5 details, unlike other disciplines that underwent profession-alization in the nineteenth century—medicine, for instance—mathematics was thus seen as grander and more efficacious at the *beginning* of the nine-teenth century than at the end. While other Victorian intellectual workers trumpeted the importance and potency of their disciplines in the march to professionalization, many mathematicians counterintuitively advanced more modest philosophies of their branch of knowledge. Intellectual de-scendents of Boole and De Morgan such as Venn and Jevons moderated the early Victorian enthusiasm that made pure mathematics and symbolic logic spiritual as well as scientific instruments. Symbols and laws that so many intellectuals had once hailed as heavenly were recast by mathematicians as earthly creations.

With the current stereotype of mathematics—dry, abstract, unrelated to larger social concerns—it is easy to forget the earlier divine proclamations about the discipline. Yet the warm-blooded sentiments behind those decla-rations form an unlikely, but critical, source out of which arose the dispas-sionate reasoning of modern philosophy and the digital logic at the heart of modern computers. And this story begins with the even earlier beliefs of an-cient philosophers such as Plato and strange occultists such as John Dee, whose writings could be found on the bookshelves of countless Victorian mathematicians.

CHAPTER ONE

Heavenly Symbols
Sources of Victorian Mathematical Idealism

I think God's thoughts after him.

—Johannes Kepler

Victorian intellectuals seeking philosophical support for a grand character-ization of mathematics did not have to look far afield. One ally was the prominent contemporary philosopher William Whewell (1794–1866), the master of Trinity College, Cambridge, in the 1840s and 1850s.[1] Although Whewell wrote on topics ranging from education, ethics, and the classics, to political economy and literature, he originally made his name in pure mathematics and associated topics in physics.[2] As a young tutor he was an early champion of the Continental system of the calculus, which blossomed among a group of his friends later known as the Cambridge Analytical Society. Whewell's most important work, *The Philosophy of the Inductive Sciences Founded Upon Their History* (1840; 2nd ed. 1847), thus unsurprisingly contained numerous laudatory chapters on pure mathematics. He proclaimed that mathematical notation, concepts, and reasoning were of such great importance that they rightly stood at the head of all "intellectual progress" in the history of mankind.[3]

Although Whewell witnessed and even participated in revolutionary advances in modern mathematics, he nevertheless thought that this lofty understanding of the discipline was as old as Western thought and believed that it was absolutely essential to understand the mathematical conceptions of certain ancient and medieval predecessors. To this end he included long digressions in *The Philosophy of the Inductive Sciences* on critical philosophers such as Plato,[4] and he sprinkled the text with exclamations from the classics that buttressed his own arguments. For instance, Whewell cited the Roman naturalist Pliny the Elder, who declared upon the mathematical prediction of an eclipse, "'Great men! elevated above the common standard of human nature, [have] discover[ed] the laws which celestial occurrences obey.'"[5] In addition to the luminaries of Greco-Roman thought, Whewell included equally long chapters on lesser-known medieval figures such as Ramon Llull.[6] Spaced throughout the history of the West, Whewell portrayed these characters as apostles who sought to spread the gospel of mathematics. The ancient Greek Plato, the Roman Pliny, and the medieval Spaniard Llull had all come to the same conclusion, though separated by the centuries—they looked upon mathematics as a discipline that communed with the highest elements of the cosmos. To them, mathematics was neither a vocation nor an avocation; rather, it was a calling.

Like Whewell, other Victorian mathematical idealists looked to these older sources for intellectual support, justification, and inspiration. Indeed, mid-nineteenth-century British and American mathematicians, especially those interested in pure mathematics, had an unusually strong sense of the history of their discipline. Simply put, they were fanatical bibliophiles. Their keen interest in the portrayal of mathematics through the ages was apparent in the rare tomes on their bookshelves and in the importance they gave to the conceptualization of mathematics above and beyond proofs and formulas. Augustus De Morgan could have been mistaken for a librarian; the list he compiled of ancient, medieval, and Renaissance mathematics books remains a useful bibliography. De Morgan also wrote biographies of premodern mathematicians for a number of encyclopedias. Similarly, John Herschel recounted the ideas of ancient Greece in a monograph for the *Cabinet Cyclopædia*.[7] George Boole and the American educator Thomas Hill (a central figure in chapter 2) both cited the Renaissance Neoplatonist John Dee as a crucial antecedent to their own thought.[8] William Rowan Hamilton went so far as to read Plato in the original Greek, and connected his own theories to Pythagorean conceptions that had influenced Plato.[9]

Because these Victorian mathematicians saw their work as the culmination of a lengthy and privileged Western intellectual lineage, the contents of their personal libraries provide an extremely useful entrée into the Anglo-American ideology of mathematics in the early nineteenth century. In these volumes, ancient, medieval, and early modern philosophers and mathematicians posited a universe divided into two planes—the sacred realm of the ideal and the profane realm of matter—with mathematics as a courier between the two. The discipline was in a unique position: available to great minds in this world, yet part of the invisible, divine sphere. Like the Victorians, many of the authors of these works yearned to transcend the fault lines of their age through mathematics.

Plato, Platonism, and Mathematics

Alfred North Whitehead may have been oversimplifying slightly when he characterized European thought as "a series of footnotes to Plato," but many of his predecessors in pure mathematics would have undoubtedly concurred. As far as we can tell, the Pythagoreans maintained a mystical belief in the power of mathematics and its symbols, but Plato and his disciples were the first to proffer a coherent, persuasive, and lasting vision of mathematics as a transcendental tool of an elite clerisy. The Victorian fathers of pure mathematics avidly read Plato, mathematically oriented Platonists such as Proclus, and British Neoplatonists of the early modern period, and they incorporated the Platonic characterization of mathematics and mathematical symbols into their own philosophies.

Although most often read as a treatise on political order, Plato's *Republic* also expounded on the special nature and potency of mathematics, and Victorian mathematicians focused on these lesser-known passages to support a transcendental definition of their discipline. Plato's advocacy of mathematics is most pronounced in Book 7, in which he described the proper education for the philosopher-kings of his ideal city. Raising the issue of exactly which topics these leaders should study to orient their minds toward the realm of the ideal Forms, Plato advocated a central role for those disciplines that "are wholly concerned with number" because "they appear to lead one to the truth."[10] Moreover, philosopher-kings should not merely dabble in the art of mathematics, but rather invest themselves wholeheartedly in the pursuit: "We should legislate and persuade those who will share the highest offices in our city to turn to arithmetic and to pursue it in no amateur spirit, but until they reach by pure thought the contemplation of the nature of numbers." Specific numbers or individual geometric shapes do not matter in and of themselves; the goal should be a higher mental state attainable through mathematics. After all, Plato's leaders employ mathematics for the greatest ends, not the mundane necessities of everyday life: "They do not pursue this study for the sake of buying and selling like merchants and retailers, but both for the sake of war and to attain ease in turning the soul itself from the world of becoming to truth and reality."[11] Thus Plato elevated mathematics to a high position in his utopia by characterizing the discipline as a "bond" (*desmos*) between this world and the ideal plane. As the character Socrates rhapsodizes about geometry to Glaucon in the *Republic*, "[It] is knowledge of that which exists forever. It would then, my noble friend, draw the soul toward truth and produce philosophic thought, and make us direct upward those parts of ourselves which we now direct downward when we should not."[12] Mathematics was a method of reasoning that resonated with the transcendental realm of Truth.

However, Victorians interested in reasserting Plato's grand estimation of mathematics, like other appropriators of Plato's authority, had to ignore certain contradictory passages in the *Republic*. Mathematics may be used to lead human beings to the ideal realm, the middle-aged Plato of the *Republic* believed, but he also cautioned against confusing the discipline with the highest participation in the Forms or the Forms themselves. For instance, in Book 6, Plato specifically placed mathematics in the *second*-highest category of human intelligence, *dianoia* (reasoning). Just above this category is the more important *noesis*, or understanding, in which human beings directly engage the realm of the Forms.[13] Since Plato stated in the *Republic* that mathematics was one of the "*preliminary* studies which must precede the dialectic,"[14] and because of his ultimate characterization of *philosophos* as higher than *mathematikos*, the idealist mathematicians of the nineteenth century had to turn to Plato's later works for more unqualified support of the transcendental conception of mathematics.

Like many intellectuals, Plato became more concerned with the spiritual as he neared the end of his life, and in his later work his proclamations regarding mathematics became less cautious and reserved. This is especially true in the *Timaeus*, with its analysis of the universe in terms of a divine and harmonious geometry. In his description of the "world-soul," Plato used mathematics extensively, showing how exact numerical principles divided the cosmos.[15] In fact, Plato believed that at every stage of the creation, from the marking of the boundaries between elements to the paths of the heavenly bodies, the Demiurge impressed upon the universe geometry and harmonic multiples.[16] To understand the true nature of the cosmos, philosophers must therefore comprehend basic mathematical principles. And when engaging in the mathematical analysis of the universe, the human soul achieves a resonance with its fundamental composition. Given this exceedingly divine conception of mathematics, a well-thumbed copy of Plato's *Timaeus* unsurprisingly could be found on the shelves of many high-minded nineteenth-century mathematicians and astronomers.[17]

An even more significant text for these Victorians, however, was a more obscure work of Plato's, the *Epinomis*. Quotations from the *Epinomis* (in Greek) often graced the title pages of grade-school math primers in the early Victorian era. Although now the *Epinomis* is only infrequently read (and frequently puzzled over), two centuries ago many intellectuals and mathematicians considered it to be an essential addendum to the all-important *Laws*—that is, the answer key to one of Plato's grandest textbooks.[18] In the twelfth and final book of the *Laws*, Plato tantalized his audience with a question posed to the members of the Nocturnal Council: How can one become supremely virtuous and satisfied? Or as he rephrased this critical inquiry at the beginning of the *Epinomis*, "What are the studies which will lead a mortal man to wisdom?"[19]

With a greater emphasis than in the *Republic*, Plato stated in the *Epinomis* that mathematics was the way to the ultimate comprehension of the universe, and that mathematical notation bridged the gap between the realm of the Forms and the physical world. Anticipating Galileo's famous dictum that God wrote the book of the universe in the language of mathematics, Plato asserted that the Demiurge "constructed us with this faculty of understanding what is shown us, and then showed us the scene He still continues to show."[20] The ability to reason mathematically was thus a sublime present from God: "With the gift of the whole number-series, so we shall assume, [the Demiurge] gives us likewise the rest of understanding and all other good things. But this is the greatest boon of all, if a man will accept his gift of number and let his mind expatiate over the whole heavenly circuit."[21] Furthermore, mathematics functions as the divine queen of the disciplines, without which all knowledge would falter: "There is every necessity for number as a foundation . . . we shall also be right if we say of the work of all the other arts which we recently enumerated, when we permitted their existence, that nothing of it all is left, all is utterly evacuated, if the art of number is destroyed."[22] In his starkest assessment of man's attempt to understand the cosmos and reach a higher state of existence, Plato claimed in the *Epinomis*, "If number were banished from mankind, we could never become wise at all."[23] One can hardly imagine a stronger assertion: Without mathematics—that God-given way of illuminating the universe—human beings would flounder in the dark.

The overriding concern with mathematics in the *Epinomis* (apparently surpassing the importance of *noesis* in his earlier work) has struck many Plato scholars as somewhat discordant with the rest of Plato's corpus; some even believe it is pseudo-Platonic.[24] This criticism seems to ignore the *Timaeus*, an accepted text, which appears to present an inchoate version of the characterization of mathematics in the *Epinomis*. However, just as the use of Bode's Law did little to taint the discovery of Neptune, the minor detail of who actually wrote the *Epinomis* mattered little to nineteenth-century mathematical idealists. They would repeatedly invoke the work to support their ideology, and would reiterate its arguments, often verbatim, in their own writings. They reveled in Plato's declaration that mathematics was a—or even *the*—divine language. Moreover, in a less high-minded way many mathematicians also concurred with Plato's belief that only a few thinkers could ascend to the transcendental state mathematical contemplation and notation promised. The "science of number" was difficult to learn and even more difficult to master, and therefore the blissful realm of profound understanding was unattainable for humanity "with the exception of a chosen few," as the *Epinomis* concluded.[25]

Plato's grander statements about the nature of mathematics perhaps would have drifted into obscurity had they not been highlighted centuries after his death by the Hellenic philosopher Proclus (ca. 410–485). Proclus wrote shrewd commentaries on Plato's mathematical works, as well as on Euclid's *Elements*. In 1789 the English philologist Thomas Taylor (a Platonist in his own right) published a translation of that latter commentary, a seminal treatise on the philosophy of mathematics, under the title *The Philosophical and Mathematical Commentaries of Proclus on the First Book of Euclid's Elements,* which became a popular reference for nineteenth-century pure mathematicians.[26] Proclus's universe, based on Plato's idealist thought, was a bifurcated one. Human beings exist in a cave of impaired perception, a profane realm of limited, imperfect things: matter, decay, ever-changing shapes. Above our muddled existence, however, is another, divine realm, a sphere of purity and eternal Truth. Mathematics plays a special role in this divided universe—it ascends from the world of impermanence to this higher, heavenly plane.

This characterization of mathematics as a courier between the sacred and the profane, clearly appropriated from Plato, became a hallmark of Procline Neoplatonism.[27] The very first line of Proclus's *Commentaries* asserted that mathematics was a unique intermediary: "Mathematical being necessarily belongs neither among the first nor among the last and least simple kinds of being, but occupies the middle ground between the partless realities [the Forms, or *Nous*] and divisible things characterized by every variety of composition and differentiation."[28] On the one hand, the highest mathematical concepts "stand in the vestibule of the primary forms, announcing their unitary and undivided and generative reality."[29] Since we human beings can use them in a practical way, however, mathematical concepts clearly "have not risen above the particularity and compositeness of ideas and the reality that belongs to likenesses."[30] In short, because of its participation in both the perfect and imperfect spheres of existence, mathematics provides a mental pathway for ascending out of the material realm and attaining an ideal comprehension of the universe. Proclus's succinct definition of mathematics as the "recollection of the eternal ideas in the soul" strikingly portrayed the ultimate purpose of the discipline.[31] At the end of the prologue to the *Commentaries,* he summarized the tremendous potency and spiritual role of mathematics:

> It arouses our innate knowledge, awakens our intellect, purges our understanding, brings to light the concepts that belong essentially to us, takes away the forgetfulness and ignorance that we have from birth, sets us free from the bonds of unreason; and all this by the favor of the god who is truly the patron of this science, who brings our intellectual endowments to light, fills everything with divine reason, moves our souls towards *Nous* [the Forms], awakens us as it were from our heavy slumber, through our searching turns us back upon ourselves, through our birthpangs perfects us, and through the discovery of pure *Nous* leads us to the blessed life.[32]

Irrationality, clouded perception, and faulty notions accumulated throughout our lifetimes obscure our vision of the truth, Proclus thought, and condemn us to the flickering shadows of Plato's cave. Thrown down from heaven, mathematics is the lightening bolt that rouses us, dispelling the darkness of the cave with an unparalleled and sublime brilliance that alerts us to the possibility of seeing everything in such a vivid, penetrating light. The mathematical arts thus involve nothing less than the purification and ascendancy of the human soul, permitting it to overcome the innate imperfection of mankind and the human senses. Placing our trust in that which lies "between absolutely indivisible realities and the divisible things that come to be in the world of matter" finally allows us to exit the cave.[33]

Beginning in the sixteenth century Procline Neoplatonism returned to prominence in Western thought.[34] In the intellectual setting of early modern Britain, mathematics became the stronghold for this idealism, as opposed to on the Continent, where Neoplatonism's main influence occurred in philosophy.[35] Victorian mathematicians looking to promote a grand vision of their discipline could therefore appeal to a robust homegrown tradition.

Preeminent among the foundational mathematical texts in the early modern period was a famous printing of the ancient work that had entranced Proclus: the 1570 English edition of Euclid's *Elements*. The book was notable not only for its fine illustrations and accomplished translation but also for its preface by John Dee (1527–1608), a prominent Elizabethan mathematician. Dee, like many heroes of later mathematicians such as George Boole, Augustus De Morgan, and Benjamin Peirce, led a colorful existence. In addition to his mathematical endeavors, he was an astrologer, alchemist, secret agent, and friend of Queen Elizabeth.

He was also perhaps the last representative of a long and influential line of Renaissance Neoplatonists. These intellectuals strove in myriad ways to discover how one could rise above the profane, physical world and commune with the divine realm of spirit. They also, of course, carefully read the later works of Plato. On the shelves of Dee's enormous private library (which was almost ten times larger than the collection at the University Library of Cambridge), he tellingly kept not one, but at least four, copies of Plato's *Timaeus*. Marsilio Ficino, the Renaissance philosopher and physician who translated Plato into Latin, summarized the legacy of works such as the *Timaeus*: "The soul contemplating the divine things assiduously and intently grows up so much on food of this kind and becomes so powerful, that it overreaches its body above what the corporeal nature can endure."[36] Ficino believed this transcendental state would lead to an absolute understanding of the universe, and thus the ability to manipulate it in magical ways.

John Dee inherited this occult quest and was convinced that mathematics was the special language that would transport its conjurer to that higher plane of divine truth. Dee's introduction to Euclid's *Elements* encapsulated the purpose and efficacy of mathematics in a manner that resonated with the mathematical idealism of the early Victorian age; Boole later called the introduction "very fruitful" and other mathematicians repeatedly cited it.[37] In a way by now familiar, Dee divided all things in the universe into three categories: the natural, the supernatural, and the mathematical. Natural things are perceivable, changeable, and divisible. Supernatural things are invisible, immutable, and indivisible. Mathematical concepts occupy a critical middle position between the natural and the supernatural, thus mediating between these realms:

> Surmountyng the imperfectió of conjecture, weenyng and opinion: and commyng short of high intellectuall cóceptió, are the Mercurial fruite of *Diancæticall* discourse, in perfect imagination subsistyng. A mervaylous newtralitie have these thinges *Mathematicall*, and also a straunge participatió betwene thinges supernaturall, immortall, intellectual, simple and indivisible: and thynges naturall, morall, sensible, compounded and divisible.[38]

Once again the fundamental definition of mathematics as simultaneously of the ideal and material spheres is apparent. Dee's conception of its "newtralitie" is also critical—mathematical analysis leads to transcendental truth untainted by the infirmities of human thought. "In Mathematicall reasonings, a probable Argument, in nothyng regarded: nor yet the testimony of sense, any whit credited: But onely a perfect demonstration, of truthes certaine, necessary, and invincible: universally and necessaryly concluded: is allowed as sufficient for an Argument exactly and purely Mathematicall," he concluded.[39] Exactly three hundred years after Dee's edition of Euclid's *Elements*, Benjamin Peirce would win acclaim among contemporaries by echoing the Elizabethan in his definition of mathematics as "the science which draws necessary conclusions."[40] For both the nineteenth-century mathematician and his sixteenth-century predecessor, the concept of "necessity" was infused with Platonic meaning—absolute declarations require communication with the pure realm of the Forms.

Further inspiring Victorian idealists such as Peirce, John Dee envisioned mathematics as more than just the path to certainty—studying it deeply was a spiritual pilgrimage as well. Pure mathematicians commune with heaven. In his preface to Euclid's *Elements*, Dee borrowed liberally from Plato's *Timaeus* in his description of the nature of the cosmos: "All thinges (which from the very first originall being of thinges, have bene framed and made) do appeare to be Formed by the reason of Numbers. For this was the principall example or patterne in the minde of the Creator." For Dee, therefore, the ability to use mathematics was a great revelatory gift from God to mankind. He optimistically declared that mathematics not only allows us to come to certain answers, but lets us "clime, ascend, and mount up (with

Speculative winges) in spirit, to behold in the Glas of Creation, the *Forme* of *Formes* . . . searchyng out and understandyng of every thyng, hable to be knowen."[41] Dee's Platonic conception of mathematics as a mediator between the natural and supernatural made the discipline essentially a religious endeavor.

In addition to Dee, the English Neoplatonist philosopher John Norris (1657–1711), a contemporary of Newton, greatly influenced the mathematical idealists of the nineteenth century. Sometimes called "the English Nicolas de Malebranche," Norris was in fact deeply indebted to that French idealist thinker, as well as to the Cambridge Neoplatonists Henry More and Ralph Cudworth.[42] Significantly, he was also the first philosopher to use Platonist arguments to attack John Locke's emphasis on the senses and the limitations of what we can know. Like Plato and John Dee, Norris broke the universe into two realms, which he called the "Ideal" and the "Natural." In his most significant work, *An Essay Towards the Theory of the Ideal or Intelligible World* (2 vols., 1701–1704), Norris asserted that the human mind, though perhaps limited, has methods to bridge these two realms and thus engage the spiritual plane. Foremost among these methods was mathematics, which Norris used as a case in point against Locke's phenomenalism. Mathematics, Norris claimed, was at heart the use of divine and eternal concepts in the analysis of the natural world; it consisted of ideas which "must have their Foundation in an Eternal Mind."[43] Indeed, mathematics lifts us up to God's "Ideal World." Norris assailed Locke's argument that mathematics relies on physical diagrams that come from the senses: "I see Mathematic Figures as they are in Idea, because I see them in such Perfection and according to such a state of Immutability as they have not in Nature . . . So that tho' the Natural World be the object of *Sense*, yet the Ideal World is the proper object of *Knowledge*."[44] Any science (in the original, broad sense of the word) that relies on mathematics therefore leads to the realm of "Eternal Essences."[45] One obvious corollary to this philosophy is that the pursuit of mathematical knowledge could be the basis of a rich spiritual life removed from cathedrals or priests. For Victorian mathematicians who felt uneasy with the Church yet who wanted to remain active in their faith, this association of religious devotion with the discipline of mathematics proved extremely appealing.

A final early modern English Neoplatonist to have a significant impact on the work of mathematical idealists in the nineteenth century was the algebraist and cofounder of the Royal Society, John Wallis (1616–1703).[46]

Augustus De Morgan had a particular interest in Wallis, owning his works in their original editions, authoring the *Penny Cyclopedia* biography of him, and highlighting the importance of the seventeenth-century mathematician to George Boole.[47] Wallis was an iconoclast like John Dee. Besides being a mathematician, logician, and grammarian, he was adept at deciphering codes and was instrumental in the decoding of Royalist letters and papers during the Civil War.

The significance of Wallis's work for the Victorians was his extension of the transcendental philosophy of mathematics into the subfield of algebra. Wallis's seminal monographs in this area, *Mathesis Universalis* (1657) and *Treatise of Algebra* (1685), reconsidered what mathematical symbols actually signify. Rather than restricting algebraic letters to their traditional role as surrogates for numbers, Wallis argued (in a manifestly Platonic fashion) that algebra was in fact a "Universal Art," a discipline with far more potency than mathematicians or others generally conceived. For Wallis, algebra was a science consisting not of mere signifiers (i.e., of numbers), but of symbols that were abstract, ideal objects.[48] "The great Advantage of Algebra," he claimed, was "that it manageth Proportions abstractly, and not as restrained to Lines, Figures or any such particular subject; yet so as to be applicable to any of these particulars as there is occasion."[49] This reconceptualization of the letters and marks used in mathematics prompted nineteenth-century pure mathematicians to investigate a wider range of uses for their discipline. In De Morgan's *Encyclopaedia Britannica* article on the history of mathematics he was not kind to Descartes or the Italian algebraists, yet he defended John Wallis's work as being of special merit.[50]

Antecedents of Mathematical Logic

One important use for Wallis's broader sense of pure algebra was in the service of linguistic clarity. Given the long Platonic tradition in the philosophy of mathematics, it should not come as a surprise that a critical field of modern pure mathematics—mathematical logic—had premodern antecedents, and the Victorians were well aware of them. Although commonplace today, the replacement of words and concepts with algebraic symbols was not a self-apparent idea. With their urge to transcend the divisions and disarray of human culture for the unity of the divine plane, medieval and early modern precursors gave the nineteenth-century founders of mathematical logic a basic conceptual framework. Their systems varied significantly from the logic of Boole or De Morgan, yet they shared a common motivation: to reconstruct the heaven-reaching Tower of Babel—this time without being thwarted by the confusion of tongues.

Ramon Llull (occasionally spelled Lull or Lully; 1232–c.1316) was the first, and one of the most frequently mentioned, progenitors of symbolic substitution.[51] For the first thirty years of Llull's life he was a pompous courtier and poet in the court of the King of Aragon, Jacob, on the Iberian peninsula. However, upon observing frightful skin lesions on the woman he adored, he gave up the life of the court and entered a monastery.[52] There, he solemnly and assiduously studied logic for the remainder of his life, arriving at a definition of the discipline that would resonate in the nineteenth century: "Logic is the art and science by means of which truth and untruth can be recognized by reason and separated from each other—the science of finding truth and eliminating falseness."[53] Yet Llull, a devout Christian and occultist, was anything but a modern secular philosopher. Logic held promise as a technique for unveiling divine truth, he believed, and thus could be a potent ecclesiastical tool. Llull conceived of his own logical system as a way for the Christian Church to approach infallibility in its understanding of the world. By extension, he also thought that his undertaking would be highly valuable in converting Jews and Muslims to the Christian faith. Llull believed that his logic could bring an end to the religious divisions of the world by appealing to a supremely efficacious, divine Christian Art.

Llull borrowed heavily from Platonic and Neoplatonic sources to create his complex system. He based his method primarily upon symbolism, and saw his notation as coming from a supernatural source, infused with the absolute power of God's domain. Llull's "Alphabet," which employed letters for specific divine attributes, embodied this idea. He substituted the letter B for "Goodness," C for "Magnitude," D for "Eternity," E for "Power," F for "Wisdom," G for "Will," H for "Strength," I for "Truth," and K for "Glory." At times he added or subtracted other symbols and letters. Llull's technique, simultaneously redolent of mathematics, astrology, and roulette, placed these symbols on rotating wheels that brought them into complex geometrical relationships with each other. The operating *magus* interpreted the resulting patterns.

Though obviously far from today's symbolic logic, some aspects of Llull's system—especially when viewed from his perspective—clearly prefigured modern efforts: the replacement of words by symbols; the manipulation of those symbols by a rigid set of rules; the acceptance of the end product of that manipulation as absolutely true because of the "rigor" of the technique. More directly significant for Victorian mathematical logicians than these perhaps tenuous links, however, was the fact that Llull's system required him to meditate on the logical meanings of "and" and "or," a critical first step toward any symbolic logic.[54] Clearly a great many differences separate Ramon Llull's *Tractatus Novus de Astronomia* from the *Principia Mathematica* of Bertrand Russell and Alfred North Whitehead. Nevertheless, as Francis Yates and others have discovered, Gottfried Leibniz had Llull's "Art" at least partly in mind when he was constructing the theoretical foundations of mathematical logic.[55] As another indication of Llull's importance, John Dee's collection of the medieval logician's manuscripts was larger than his stock of any other author's.[56]

Early modern British mathematicians were somewhat less ambitious than Llull, yet they too yearned for a transcendental symbolic technique. One central figure in this quest was Thomas Harriot (1560–1621), a precursor often on the lips and in the minds of nineteenth-century pure mathematicians. Harriot was a pioneer in many respects: a geographer on the second expedition to Virginia in 1585, a religious nonconformist who denied the divinity of the Old Testament, and a mathematical innovator of the first order. He believed fervently in the existence of imaginary numbers, a bold position for his age, and was responsible for creating the comparison symbols ">" and "<". Like Llull, Harriot created a rudimentary symbolic system based on a set of fundamental laws. Utilizing algebraic notation he tried to create a new phonetic alphabet that would, in a sense, express all languages. Although some scholars have detected a Cabalistic streak in Harriot's project,[57] his goal seems to have been decidedly less mystical; Harriot was mostly interested in using his system to record and decipher the languages of New World natives.[58] Regardless of his actual intention, Victorian mathematicians saw Thomas Harriot as another prominent forefather of the reduction of language to a mathematical format, and it was this particular element of his wide-ranging work that fascinated them. Augustus De Morgan was especially intrigued by Harriot's system (and his liberal theology), writing about him to interested colleagues such as John Herschel.[59]

The early modern system most germane to the mathematical logic created in the Victorian age was the symbolic theory of the mathematician and Bishop of Chester, John Wilkins (1614–1672). Wilkins, a founding member of the Royal Society, was another historical favorite of nineteenth-century pure mathematicians. In his monograph *An Essay Towards a Real Character, and a Philosophical Language* (1668), he attempted to create a new dialect based on symbols that would rise above the uncertainties and inconsistencies of normal speech. In a phrase strikingly reminiscent of the work of the Victorian mathematical logicians, Wilkins sought "the distinct expression of all things and notions that fall under discourse."[60] His contemporary Thomas Hobbes had tried to do much the same thing in the first book of his *Leviathan* (1651), but Wilkins's project was far greater in scope and more removed from words themselves.

Profoundly distraught that the manifold languages of humanity contained innate deficiencies and generated widespread confusion, Wilkins believed that he could reduce them all to a set of basic laws. "Abstracting from many unnecessary rules belonging to instituted languages," as he put it, he set about constructing a series of tables that would encapsulate the fundamental relationships used in communication.[61] From these tables Wilkins then went about the Herculean task of distilling the entire dictionary into a succinct format, hoping to arrive at a kind of *ur*-language. Ultimately, he thought he could transcend the long, inefficient words of the vernacular with an ideal language consisting of an assortment of marks similar to those used in mathematics. These symbols, Wilkins believed, were far clearer than words: "If every thing and notion had a distinct mark, with some provision to express grammatical derivations and inflections, it would answer one great end of a real character, to signify things and not words."[62] For Wilkins, symbols were akin to crystalline Platonic Forms; unlike words, they were not multivalent and thus they provided a secure foundation for rational discussion and deliberation.

The Victorians who founded modern mathematical logic may have judged John Wilkins's system defective, but they clearly emulated his basic idea and strongly agreed with the motivation behind it. They envisioned their century like the seventeenth century—an era with a high potential for sectarian strife. And like Wilkins, mathematicians such as Boole and De Morgan sought ways to rise above such divisiveness. This ecumenical streak ran through the technical marks and tables of *An Essay Towards a Real Character, and a Philosophical Language*. "The advantages proposed by this philosophical language," Wilkins wrote, "[are] the facilitating of mutual commerce among the several nations of the world; the improving of natural knowledge; and the propagation of religion." In a passage that would resonate in the nineteenth century, Wilkins prophesied the broad application and power of his method, especially in the service of faith:

> It might contribute much to the clearing of some modern differences in religion, by unmasking many wild errors that shelter themselves under the disguise of affected phrases: which being philosophically unfolded, and rendered according to the genuine and natural importance of words, would appear to be inconsistencies and contradictions; and several of these pretended mysterious profound notions, expressed in big swelling words, by which men set up for reputation, being this way examined, would either appear to be nonsense, or very jejune.[63]

Words can lie, distract, mislead, confuse, and advance false religion; clean notation, exemplified in the symbols of mathematics, cannot.

Although they worked in a putatively technical and forward-looking discipline, by looking back to figures such as John Wilkins, early Victorian mathematical logicians saw themselves as providing a means for the resolution of seemingly intractable cultural and religious problems. Nineteenth-century mathematical logic, like Wilkins's algebraic dialect of the seventeenth century, would, in Wilkins's words, hopefully "prove the shortest and plainest way for the attainment of real knowledge, that had yet been offered to the world."[64] A new language, based on the ideal nature of mathematics, would purify and elevate the profane world.

Indeed, the power to rise above the mire of innumerable languages and opinions became a potent theme in pure mathematical research in the Victorian era. As John Herschel lamented just before George Boole and Augustus De Morgan began to develop this new logic, "It is, in fact, in [the] double or incomplete sense of words that we must look for the origin of a very large portion of the errors into which we fall."[65] What was the best way out of this predicament? Herschel, like Plato, had a clear idea of the direction to take:

> The study of the abstract sciences, such as arithmetic, geometry, algebra . . . being free from these sources of error and mistake, accustom us to the strict use of language as an instrument of reason, and by familiarising us in our progress towards truth to walk uprightly and straight-forward on firm ground, give us that proper and dignified car-

riage of mind which could never be acquired by having always to pick our steps among obstructions and loose fragments, or to steady them in the reeling tempest of conflicting meanings.[66]

Simplification and clarification of discourse through the transcendental reasoning and symbols of mathematics appeared to be a traditionally sanctioned solution to a difficult problem. As the poet William Wordsworth put it, his contemporaries must condemn "that false secondary power/by which we multiply distinctions, then/Deem that our puny boundaries are things/ That we perceive, and not that we have made."[67] The fathers of mathematical logic followed Herschel and Wordsworth in their search for truth that originates above such groundless human classifications.

Modern Idealism and Romanticism

Ancient, medieval, and early modern thinkers who advanced mathematical Platonism might have remained historical curiosities were it not for the boost given to philosophical idealism in the late eighteenth and early nineteenth centuries. The decades following the French Revolution were a time of political, cultural, and religious turmoil that rekindled a strong interest in transcendental notions among intellectuals. By describing the universe as split between the "phenomenal" and "noumenal" realms—that is, between the realm of physical matter, space, and time, and the atemporal and nonspatial realm inhabited by "things-in-themselves," God, and the human soul—Immanuel Kant, in particular, inaugurated a robust period of philosophical idealism in Europe and America. Kant's philosophy naturally engendered a strong desire to understand how human beings could access the elusive noumenal realm. He postulated that this communion occurred through a special faculty of the mind—a transcendental form of reason, as opposed to the speculative reason that analyzes the material realm. Filtered through the British Romantics, Kant's twin concepts of the noumenal realm and transcendental reason provided a theoretical framework for the lofty construction of pure mathematics in the Victorian era.

Finding support for the divinity of mathematics in Kant's philosophy required a myopic and somewhat distorted view of his thought, however.[68] It all depended on which part of his definition of mathematics one chose to emphasize. Kant held that mathematics was "synthetic a priori" knowledge;

"synthetic" meaning it could add to our storehouse of information, and "a priori" indicating it did not come from the experience of the senses. When Kant expanded on the first part of the definition, "synthetic," he seemed hesitant to characterize mathematics as genuinely transcendental. In these passages he indicated that pure mathematics was humanly constructed, rather than eternally present in the noumenal realm. In the *Critique of Pure Reason*, for example, Kant concluded that higher-order mathematical laws "contain an arbitrary synthesis."[69] We establish basic mathematical definitions and rules in our mind and then proceed to combine them to generate more complex definitions and rules. Although mathematics does not have its origin in the physical world, Kant appeared to be arguing, its laws do not originate in the transcendental realm either.

Nevertheless, Victorian mathematical idealists lauded Kant for removing their discipline from the province of mere speculative reason based on the impressions of the senses. As the tide of Cambridge Neoplatonism ebbed in the late seventeenth century, many seminal British thinkers, including John Locke and David Hume, came to perceive mathematics as a refinement of everyday sensory analysis. All mathematical facts were a posteriori knowledge, gained from mundane tasks such as the measuring of physical objects. Kant's fervent belief that mathematics was a priori knowledge, in contradistinction to such materialistic characterizations, thus carried significant weight. The mathematical law that parallel lines never meet is impossible to attain from experience (since lines are infinite in length), Kant noted, yet it is an important fact of (Euclidean) geometry, and it leads to other profound insights. With examples such as this, Kant used mathematics to highlight how human beings could acquire grand ideas without the senses.

Furthermore, Kantian idealism meshed well with older Platonic sources, combining to give the discipline of mathematics an air of clairvoyance. Thinkers such as Proclus, Dee, Norris, and Wallis told Victorian mathematical idealists what many wanted to hear in concert with Kant's philosophy: mathematics was the great mental faculty that would allow human beings to gain access to the invisible kingdom of Absolutes. Just as Kant's cautious statements about the noumenal realm were ignored by countless idealists in the nineteenth century, so were his less grandiose passages about mathematics ignored by those interested in giving the discipline a heavenly pedigree. It would take just a few small nudges to move mathematics entirely into the noumenal realm.

Samuel Taylor Coleridge (1772–1834), the Romantic poet and wholesale importer of German philosophy into the English-speaking world, was one of those who prodded the philosophy of mathematics in the direction of the divine. His strong Neoplatonic streak, along with his reading of Kant, Schelling, and the *Naturphilosophen,* helped push Anglo-American thought away from its focus on the world of matter.[70] Like John Norris, Coleridge wished to salvage a higher view of the universe from contemporaries who, he believed, improperly concentrated on the base and particular. By highlighting the ability of the human mind to transcend the deterministic causal chain of the phenomenal realm, Coleridge was largely responsible for reinvigorating philosophical idealism at the beginning of the Victorian age. Furthermore, while he spurred countless English-speaking writers and artists to seek the noumenal realm, his influence was not restricted to those in the humanities. Mathematicians, too, became more interested in the ideal realm after reading Coleridge.

Appropriating Kant's conception of transcendental reason, Coleridge argued that lofty ideas, not matter, should be at the center of any true system of philosophy. Foremost on Coleridge's list of philosophers who were guilty of the converse was Spinoza, whose notion of God as "substance" made him the *bête noire* of generations of spiritual European intellectuals, including George Boole and other mathematical idealists in the nineteenth century.[71] In opposition to Spinoza's brand of philosophy, Coleridge asserted that there were "organs of spirit . . . framed for a correspondent world of spirit," that is, faculties of the mind that connect with the noumenal realm.[72] Similarly, he sought to debunk mechanistic materialism, the theory that the world is nothing but matter in motion. In his *Aids to Reflection* (1825), Coleridge identified this threatening philosophy that had arisen out of the Scientific Revolution: "It is exclusively to sensible Objects, to Bodies, to modifications of Matter, that [the materialist] habitually attaches the attributes of reality, of substance. Real and Tangible, Substantial and Material, are Synonymes for him."[73] Coleridge then posed deflating questions for the adherents of this corpuscular philosophy, including the query most crucial for idealists: How does the mind arise out of inert matter?[74] Just as it is ridiculous to expect "the paper, ink, and differently combined straight and curved lines of an Edition of Homer to bear what we understand by the words, Iliad and Odyssey," so is it equally absurd to look to atoms bumping into one another to explain consciousness or life.[75] As Coleridge had declared in an earlier work, matter "could only engender something homogeneous with itself. Motion could only propagate motion. Matter has no *In-*

ward."[76] He bristled at how materialists could not even begin to explain the true wonders of the universe: "And how came the *percipient* here? And what is become of the wonder-promising Matter, that was to perform all these marvels by force of mere figure, weight, and motion?"[77] Clearly something higher was at work in the cosmos—we need only look into our own minds to comprehend that essential truth.

Not content simply to criticize the materialist *Weltanschauung,* Coleridge offered a new vision of science informed by idealism.[78] He did so by emphasizing the supreme significance of timeless laws instead of day-to-day observations: "In whatever science the relation of the parts to each other and to the whole is predetermined by a truth originating in the *mind,* there we affirm the presence of a law."[79] For Coleridge, such laws were not mere generalizations of patterns found in the universe, but rather transcendental ideas that determined those patterns. In his "Essays on Method" in *The Friend,* he highlighted this characterization of scientific discoveries: "Let it not be forgotten . . . if they do not excite some master IDEA; if they do not lead to some LAW . . . the discoveries may remain for ages limited in their uses, insecure and unproductive."[80] Physical data ultimately must lead to the unveiling of some higher, abstract concept, achieving congruence with what Coleridge considered the divine Mind.[81] As Trevor Levere has summarized Coleridge's vision of science, "[He] was passionately convinced that the world was not made up of disconnected fragments . . . Nature's laws, originating in the mind, were god given, and part of the unity of nature."[82] For Coleridge, idealism therefore offered salvation from troubling modern philosophies by illuminating a universe in which both human consciousness and the external world provided evidence of transcendental divinity.

Since scientists expressed their highest laws in the language of mathematics, Coleridge in turn placed that discipline on a lofty pedestal. Unfortunately, he did not possess a great deal of mathematical knowledge. A telling lament from one of Coleridge's notebooks shows how tormented he was about his failure to immerse himself successfully in this critical discipline, even though his teachers included luminaries such as the astronomer and navigator William Wales:

> O! with what bitter regret, and in the conscience of such glorious opportunities, both at School under the famous Mathematician, Wales, the companion of Cook in his circumnavigation, and at Jesus College, Cambridge, under an excellent Mathematical Tutor, Newton all *ne-*

glected, with still greater *remorse!* O be assured, my dear Sons! that Pythagoras, Plato, Speusippus, had abundant reason for excluding from all philosophy and theology not merely practical those who were ignorant of Mathematics.[83]

Pure mathematics was a necessary precursor to all higher regions of thought, Coleridge believed—a sore oversight on his part, indeed. Nevertheless, he studied Proclus, including Thomas Taylor's translation of *The Philosophical and Mathematical Commentaries on the First Book of Euclid,*[84] and such reading made him appreciate the power of the symbols of pure mathematics. Later, Coleridge would proclaim in *Biographia Literaria* that "an idea, in the *highest* sense of that word, cannot be conveyed but by a symbol."[85] Moreover, symbols that point to the Eternal reside deep within our minds: "In looking at objects of Nature while I am thinking . . . I seem rather to be seeking . . . a symbolic language for something within me that already and forever exists, than observing anything new."[86] Victorian mathematicians who envisioned themselves summoning grand, ideal laws through their algebraic systems, rather than creating these laws from scratch, understood well what Coleridge meant.

Furthermore, Coleridge shared with the founders of mathematical logic a hatred of nebulous language. In his *Aids to Reflection,* an aside about semiotic imprecision demonstrated this concern: "I have neglected no occasion of enforcing the maxim, that to expose a sophism and to detect the equivocal or double meaning of a word is, in the great majority of cases, one and the same thing."[87] Like the mathematical logicians who followed him, Coleridge ultimately believed that "all true science is contained in the Lore of Symbols & Correspondences," in a realm where the confusion of everyday language simply does not exist.[88] Although individuals and groups exploit the many meanings of words to promote their viewpoints, in this higher realm only propositions "wholly independent of the will" endure. This neutrality is crucial—a properly formed science should help to establish true religion, rather than antagonize it. When science "remains neutral," when it does not try to dispel faith or offer its own disenchanted "religion," Coleridge concluded, it "becomes an effective ally [of faith] by exposing the false shew of demonstration."[89] George Boole, among others, would use the "unbiased" symbols of mathematics in precisely this way.

In his widely overlooked work *Logic* (probably finished in the 1820s), Coleridge crystallized his ideas regarding the nature and prominence of mathematics.[90] A full three chapters of the manuscript, and a significant portion of the remainder, concerns mathematics, an important focus since Coleridge understood that his case for idealism turned on a single issue: convincing his audience of the existence of objects that are immaterial yet certain. Like Kant (whose work he often creatively plagiarized), Coleridge addressed the skepticism of modern philosophers such as David Hume by appealing to knowledge that seemed to be innate to the mind, existing independently of the external world.[91] Hume believed that mathematics was analytical—that is, based on our observations of the world. For example, we know that $2 + 3 = 5$ because we have previously put together 2 apples and 3 apples and gotten 5 apples. In *Logic,* Coleridge strenuously objected to Hume's characterization of mathematics, and he countered with a penetrating question: how can we know the sum of two immense numbers, for example, 35,942,768,412 and 57,843,647, given that we could never count that many apples even with an inexhaustible orchard and decades of free time? Instead, mathematics must consist of pure laws manipulated in the *mind's* eye. In other words, mathematics is synthetic a priori knowledge.[92] "We need only ask ourselves whether . . . we could ever have arrived at the properties of the cycloid or the proportion of the area of the curve to the area of the generating circle" using mere observation, to understand the purely ideal nature of mathematics, Coleridge concluded.[93]

From piercing points such as these, mathematics widened into a philosophical wedge. If we can show that such a priori knowledge exists, Coleridge insisted, we should not be hesitant to accept that additional a priori knowledge is possible, including supreme notions such as the existence of God and the perpetuation of the soul beyond death. Following his three chapters on mathematics, Coleridge thus proceeded to show that other a priori knowledge is indeed attainable—even knowledge of the transcendental realm. Because so much of his argument relied on the case of mathematics, Coleridge's view of the discipline was commensurately grand. For instance, at one point in *Logic* he preached about mathematics, "We have only to attempt raising our minds to a comprehension of the mighty pile and fabric of truth, which (faith in God and the moral law alone excepted) is the proudest honour and glory of the human intellect, that in which above all others it finds the clearest sense of its own permanency and at the same time the most infallible evidence of its progressiveness."[94] For Coleridge, therefore, pure mathematics would serve as a bridge between the material and divine realms, between skepticism and faith. Perhaps the most telling passage in *Logic* comes not from Coleridge himself but from the eighteenth-century poet Mark Akenside, whose revelatory stanzas about mathematics Coleridge quoted at length:

> Such is the rise of forms
> Sequestered far from sense and every spot
> Peculiar in the realms of space and time;
> Such is the throne which man for truth amid
> The paths of mutability hath built,
> Secure, unshaken still; and whence he views
> In matter's mouldering structures, the pure forms
> Of triangle or circle, cube or cone,
> Impassive all; whose attributes nor force
> Nor fate can alter. There he first conceives
> True being, and the intellectual world
> The same this hour and ever. Thence he deems
> Of his own lot; above the painted shapes
> That fleeting move o'er this terrestrial scene
> Looks up; beyond the adamantine gates
> Of death expatiates; as his birthright claims
> Inheritance of all the works of God;
> Prepares for endless time his plan of life,
> And counts the universe itself his home.[95]

Human beings first comprehend the wondrous connection between the human mind and the divine Mind when they encounter the pure forms of mathematics, and are then set on the path to understanding the existence of heaven and the nature of God. Seeing mathematical symbols as transcendental thus placed Coleridge among the most important supporters of pure mathematics in the early Victorian era.

Another central figure of British Romanticism, William Wordsworth (1770–1850), also maintained exceedingly favorable views of mathematics. Although like Coleridge he knew relatively little about the discipline, Wordsworth nonetheless advocated a sublime conception of it. Autobiographically describing his flirtation with "the geometric science," Wordsworth exalted,

> With Indian awe and wonder, ignorance
> Which even was cherish'd, did I meditate

Upon the alliance of those simple, pure
Proportions and relations with the frame
And laws of Nature, how they would become
Herein a leader to the human mind.[96]

The purity and force of mathematics clearly entranced the young poet, and it played a critical role in his intellectual growth. Mathematics, Wordsworth claimed, helped to lift him out of this profane world into the divine realm:

From this source more frequently I drew
A pleasure calm and deeper, a still sense
Of permanent and universal sway
And paramount endowment in the mind,
An image not unworthy of the one
Surpassing Life, which out of space and time,
Nor touched by welterings of passion, is
And hath the name of God.[97]

Wordsworth, echoing Akenside's declaration about the role of mathematics, obviously inherited the Platonic understanding of the discipline as a spiritual endeavor that elevated the human mind. Although a poet, in his own way Wordsworth chased the same eternal laws sought by idealist mathematicians and scientists in the Victorian age. In his poem *Excursion,* for instance, Wordsworth pursued an ideal entity

That is the visible quality and shape
And image of right reason; that matures
Her processes by steadfast laws; gives birth
To no impatient or fallacious hopes,
No heat of passion or excessive zeal,
No vain conceits; provokes to no quick turns
Of self-applauding intellect; but trains
To meekness and exalts by humble faith;
Holds up before the mind, intoxicate
With present objects and the busy dance
Of things that pass away, a temperate show
Of objects that endure.[98]

In his search for such transcendental objects, Wordsworth furthered the tradition of British Neoplatonism, and in turn contributed to the mind-set of many nineteenth-century idealists.

Idealist notions from seminal Romantic figures such as Wordsworth and Coleridge affected a range of intellectuals beyond the creative arts. Mathematicians like George Boole and Benjamin Peirce also felt the immense draw of the Eternal realm conjured by these poets. In some cases their influence was quite direct. For example, Coleridge was a good friend of William Rowan Hamilton, and he sent the eminent Irish mathematician his own valued copy of Kant's *Critique of Judgment*.[99] In turn, Hamilton closely followed the development of Coleridge's own philosophy and criticism, going so far as to copy sections of Coleridge's *Aids to Reflection* and *The Friend* into his notebooks.[100] Hamilton was also friendly with Wordsworth, concurring with his opinion that the best science rose above mere observation into a meditative unity with God's mind.[101] As the mathematician had written when he was twenty years old and in the midst of formulating his mature philosophy, "I know that Science presents to its votaries some of the sublimest objects of human contemplation; that its results are eternal and immutable verities; that it seems to penetrate the counsels of Creation, and soar above the weakness of humanity."[102] Victorian mathematicians such as Hamilton agreed with Wordsworth and Coleridge in claiming that truth was divine and that it existed in the form of heavenly symbols and laws. Combined with the older Platonist tradition, the religious idealism of the Romantic movement thus generated a school of thought in Britain and America that treasured mathematics as the way to ascend to the realm of spirit.

Descriptions of the 1846 discovery of Neptune highlighted how Romanticism helped Victorian mathematical idealists translate their awe about the planetary unveiling into words. A gleeful passage about the mathematician's work from J. P. Nichol's *The Planet Neptune* furnishes an apt example, simultaneously recalling Wordsworth's odes to the ideal and the Romantic painter Caspar David Friedrich's iconic painting of a solitary figure on a mountain top above the clouds:

Am I indeed overcharging it, in deeming that the attitude of the Inquirer here approaches the Sublime? Standing on the summit of a pinnacle to which the loftiest minds had heretofore looked with rather an aspiration than a hope, his first glance is even farther onwards,—his thoughts stretch towards remoter Altitudes still lying cloud-capped, but which may one day be scaled, and the perspective beneath them spread before the triumphant eye of Man![103]

For Nichol, as for so many other intellectuals influenced by Romanticism, mathematics clearly transported one to a transcendental place.

Religious Purity and Pure Mathematics

Ancient Platonist, early modern Neoplatonist, and Romantic testaments to the lofty character of mathematics united with a widespread nineteenth-century desire to return to a primeval faith. It is perhaps a truism to say that since the Protestant Reformation most theological reformers have considered themselves to be purifying religious ideas and practice, yet purification is an apt way to describe the synthesis of mathematics and faith in the Victorian age. By turning to symbols they believed came from heaven, religious mathematicians felt they could rise above the insecurity and chaotic diversity of human culture. Faced with an age they saw as corrupt and defective, these intellectuals sought to use pure mathematics to cleanse their muddied world. In an odd way, this devotion to the simplicity and transcendence of pure mathematics was thus kindred with the Gothic revival associated with Britons such as John Ruskin and A. W. Pugin, with elements of the Second Great Awakening in the United States, and with many other spiritual and spiritually tinged movements that proliferated in the nineteenth century in response to the perceived threat of a secular modern world and the uninspired religiosity of the established churches.[104]

To use an older comparison, in Augustine's sense mathematical idealists in the Victorian era saw themselves as a new City of God, a community that spoke a heavenly dialect and was dedicated to the one true faith that spanned the centuries and the barriers of language and culture. George Boole, for example, believed that "the progress of knowledge and the arts . . . forms a bond which connects the different generations of men together by interests and feelings wider than those which are merely national."[105] Indeed, Boole often spoke like Augustine, a model for living devoutly in a profane age. The mathematician referred to "the people of the unseen God" and how "in all ages it had been their duty to keep themselves from too close contact with idolatry."[106] In a series of emotional poems written at the same time as his initial work on mathematical logic, Boole honored this timeless brotherhood of learned believers. He was certainly no Wordsworth or Coleridge, but Boole's poetry shows that he was equally moved by sublime truth. One poem, dedicated to his mentors, began with a search for the "city" of faithful teachers:

> Fellowship of spirits bright
> Crown'd with laurel, clad with light
> From what labours are ye sped,
> By what common impulse led,
> With what deep remembrance bound,
> From the mighty concourse round
> Do ye thus together stand,
> An inseparable band?[107]

These noble compatriots, Boole believed, were to be found "beyond time and place"—a college that stretched across the ages and the continents. Included in this fellowship were those who fought for justice and ecumenical unity:

> All who felt the sacred flame
> Arising at oppression's name
> All who toil'd for equal laws
> All who lov'd the righteous cause
> All whose world-embracing span
> Bound to them each brother man.[108]

The mathematician did not merely honor religious leaders. Equivalent positions in his heavenly city were reserved for the Copernicuses and Newtons:

> All who with a pure intent
> Were on Nature's knowledge bent
> Watch'd the comet's wheeling flight
> Trac'd the subtle web of light
> And the wide dominion saw
> Of the Universal Law.[109]

Boole considered all pursuers of God's divine truths, whether ethical or scientific, as part of a timeless effort to unify the faithful and advance God's dominion. His declarations thus truly restated the main themes of Augustine's *City of God*, and Boole concluded his poem with two stanzas that could have been written by the Church Father:

> If with pure and humble thought
> For the Good alone they wrought,
> When the earthly life is done
> In the heavenly they are one.
>
> And their souls together twine
> In a Fellowship divine,
> And they see the ages roll
> Onward to their destin'd goal,
> Dark with shadows of the past
> Till the morning come at last
> And an Eden bloom again
> For the weary sons of men.[110]

Divine truth, Boole proclaimed, united its advocates beyond the ages and petty cultural distinctions. Mathematicians therefore had a chance to become a part of the glorious "Fellowship divine," to work in the service of a higher purpose. In this sense the pure mathematical and logical revolution of the nineteenth century could be a spiritual quest as well as a technical one. Boole and many other Victorian mathematicians labored in the belief that they were spreading the ideas of God to the brotherhood of humanity.

Despite such resonance with centuries-old Christian themes, however, Victorian mathematical idealists diverged significantly from theologians like Augustine. Although most were nominally Christians and believed in the ethical message of Jesus, the idea of communing with a unified divine Mind naturally pushed them toward a denial of the Trinity and a skepticism about the importance of the Church and the clergy. Unsurprisingly, many nineteenth-century pure mathematicians strayed far from orthodoxy, drifting especially toward Unitarianism. Augustus De Morgan was a prominent Unitarian, as was Benjamin Peirce; George Boole, though fearful of declaring his faith publicly, was closest in spirit to Unitarianism and cherished the work of the American Unitarian William Ellery Channing above virtually all other theologians. This trio of leading pure mathematicians thought that the "Fellowship divine" was not to be found in traditional ecclesiastical institutions but in an upstart denomination they found to have a strong idealist and ecumenical bent. Indeed, the mixture of Unitarianism and mathematical idealism was a powerful undercurrent to research in pure mathematics and mathematical logic in the Victorian age. Peirce, born within the broad theological latitude of the United States and unrestrained in his religious proclamations, most clearly exhibits this important bond between nineteenth-century Unitarian theology and pure mathematics, between religious idealism and the "heavenly symbols."

CHAPTER TWO

God and Math at Harvard
Benjamin Peirce and the Divinity of Mathematics

The Victorian congruence of pure mathematics and religious idealism found its greatest advocate in perhaps the most prominent American mathematician of the nineteenth century: Benjamin Peirce. Peirce, generally known today only as the father of Charles Sanders Peirce, the seminal philosopher of Pragmatism, was internationally famous and influential himself. Holding a position at Harvard for a pivotal half-century in which the university shifted its character away from clerical training toward a new research model, he trained and guided the work of two generations of mathematicians who would go on to train and guide countless others. Personally, Peirce was responsible for numerous theories and applications, and his *Linear Associative Algebra* (1870) was a landmark both for its new form of algebra and for its statements regarding the nature of mathematics itself. A close friend of many important religious thinkers of the time, including Ralph Waldo Emerson and founding theologians of American Unitarianism, Peirce transferred their spiritual notions into the realm of mathematics. In many ways Benjamin Peirce, often called "the father of pure mathematics in America,"[1] epitomized the symbiosis of faith and mathematics in the Victorian age: fervent ecumenist, vanguard educational reformer, and unrepentant elitist. In concert with a small but significant band of fellow travelers at Harvard, especially the Unitarian clergyman and university administrator Thomas Hill, Peirce developed a sense of mathematics that was inseparable from his positions on the heated issues of the day.

At Peirce's funeral Andrew Peabody, a mathematician and professor of Christian morals at Harvard, highlighted their common religious conception of mathematics. He wrote that Peirce, though not a clergyman, knew "more about the realm of spiritual being than any one else who ever trod the earth, that he beheld God, entered into the Divine mind, drank in truth from its living and eternal fountain, as no other human being ever did." Peirce spent his life, in Peabody's words, in "the vivid, eager pursuit of the eternal truth of God, of which the signs and quantities of mathematics are the symbols . . . In [his] lectures he has shown, as he always felt with adoring awe, that the mathematician enters as none else can into the intimate thought of God, sees things precisely as they are seen by the Infinite Mind, holds the scale and compasses with which the Eternal Wisdom built the earth and meted out the heavens."[2] This bond between his vocation and his faith was the hallmark of Peirce's thought, a mathematical spirituality with as much intensity and sincerity as any nineteenth-century revival. His theology de-emphasized the core dogmas of Christianity and indeed the figure of Christ himself, settling instead on a broad monotheistic faith in which the quest for mathematical truth and the quest to know God were identical. Benjamin Peirce saw his work with equations as a way to access the heavenly realm, and would occasionally add the exclamation "Gentlemen, there must be a God" to his mathematical demonstrations.[3] The notion that mathematics was a divine language suggested to Peirce that sectarian distinctions were artificial and thus not to be tolerated, that advanced research in abstract fields such as mathematics should be the main priority of a university rather than teaching, and that, although all men were created in the image of God, some individuals attained a closer kinship with His mind.

Peirce grew up in Salem, Massachusetts, a contemporary of the town's most famous son, the writer Nathaniel Hawthorne.[4] Although relatively small, Salem was home to a disproportionate share of mathematicians and scientists, including the physician Edward Augustus Holyoke (1728–1829) and the mathematician Nathaniel Bowditch (1773–1838). Bowditch was the premier American mathematician of his age, an innovator of the first order who imported European methodologies and notation. Just as important, a series of outstanding libraries were located in Salem. From 1760 onward, residents had access to the Social Library, the Kirwin Library, the Philosophical Library (instrumental in the education of Bowditch), and finally the consolidated Salem Athenaeum.[5] The intellectual culture of the city and its research collections provided the young Peirce with great resources, as did his parents. His father, also named Benjamin, married his cousin Lydia Ropes Nichols at the turn of the century, and became a merchant for three decades with occasional stints as a state legislator. However, in midlife the senior Benjamin Peirce changed careers and became the librarian and historian of Harvard College.

A year earlier the younger Benjamin Peirce had set off for Harvard himself, arriving on campus in 1825 just as Unitarians ascended to a dominant position at the college.[6] The entering Class of 1829 included a number of religious nonconformists who would later achieve prominence in America: the writer Oliver Wendell Holmes (Sr.), the Unitarian clergymen William Henry Channing and James Freeman Clarke, and the Supreme Court justice Benjamin R. Curtis. Nicknamed "Benny" by these classmates, Peirce had an oversized head and imposing brow, and he cultivated the kind of bushy-bearded, unkempt look that bespoke obsessive erudition. He devoured volumes on pure mathematics the way his friends consumed novels and poetry.[7] The self-tutelage paid off—in his freshman year he topped all other undergraduates to win the highest mathematical honors of the university.[8] By the age of twenty, Peirce was helping Nathaniel Bowditch translate and annotate Laplace's revolutionary astro-mathematical work, *Mécanique Céleste*, and a couple of years after graduation he secured a position at Harvard as a tutor. A mere year later he became the Hollis professor of Mathematics and Natural Philosophy, and a decade after that added a chair in astronomy. In those capacities Peirce wrote on facets of pure mathematics—trigonometry, geometry, algebra, the calculus—as well as on topics in mechanics, the motion of stars, fluid dynamics, and even theoretical meteorology. Beyond his academic duties, Peirce became the consulting astronomer to the *American Ephemeris and Nautical Almanac,* the President of the American Academy of Arts and Sciences, and the Superintendent of the United States Coast Survey. At Harvard he also (like his father) served a stint as the college librarian, and by lecturing on a comet in 1843 helped secure funding and support to build the Harvard Observatory. Peirce's preliminary calculations regarding irregularities in the orbit of Uranus were instrumental in the work of Urbain Le Verrier and John Couch Adams. In the final decade of his life, following the publication of *Linear Associative Algebra,* Peirce retreated from his campus celebrity. He handed over the reins of his professorship to his son James and gave himself wholeheartedly to theology and philosophy, elements that had simmered beneath his work in mathematics. Religious concerns formed the core of what he believed to be his true magnum opus, the two-hundred-page *Ideality in the Physical Sciences,* published posthumously in 1881.

Unitarianism and Transcendentalism

While his public image would seem to have entailed a life of solitary number crunching, Peirce thrived on discussion with other intellectuals, most of whom were like-minded religious idealists. Throughout his life he was particularly affected by his commerce with Ralph Waldo Emerson and Unitarian theologians. The transcendentalist Emerson, of course, was forthrightly a promoter of idealism; the Unitarian clergymen Peirce admired and associated with in his formative years were equally so disposed. Indeed, Emerson came to see the Unitarianism of the 1840s as a religious incarnation of philosophical idealism.[9] The Platonist bent of Emerson and the mid-nineteenth-century American Unitarians—focused on accessing a higher plane rather than scriptural, liturgical, or ecclesiastical concerns—matched Peirce's priorities. From the perspective of these idealists, God cast the human mind in the mold of His divine Mind, which allowed for a spiritual understanding of pure mathematical thought. Religious sources therefore helped Peirce to discover the meaning and motivation behind his work with variables and formulas. Not only was mathematics a profound spiritual endeavor in this idealist vision, but it also held the potential to reinvigorate faith in a way that undermined the authority of the mainstream clergy and churches.

This affinity between Unitarian theology and pure mathematics comes into view if one looks at the nature of Unitarianism in the first half of the nineteenth century. The word Unitarian was originally a derogatory term, used in the eighteenth century by its opponents as a rebuke of those who failed to acknowledge the doctrine of the Trinity.[10] Early Unitarians such as Richard Price saw things differently, of course, believing they were returning to an original Christian faith unfettered by dogma, including the Athanasian Creed, that had been developed long after the life of Jesus. In the early nineteenth century, however, Unitarian theologians shifted their concern from the exact nature of godhood to a greater concentration on the relationship between man and God. This latter topic was far more contentious. By the beginning of the Victorian age, in fact, Unitarianism was anything but unified. One branch of the sect looked to the Bible for piety and Scottish realism for philosophy, and remained a coherent church. Another branch of Unitarians found their inspiration in Kant and German idealism instead.[11] "I have long seen that the Unitarians must break into two schools,—the Old one, or English school, belonging to the sensual and empiric philosophy,—and the New one, or the German school (perhaps it may be called), belonging to the spiritual philosophy," the Unitarian minister and Harvard Divinity School professor Convers Francis wrote in his diary in 1836.[12] The "German school" argued for communion with God through Kant's transcendental reason. They derided Lockean philosophy, with its emphasis on sensory experience, in favor of the inner intuition of religious truth.[13] The theological tenets and antiecclesiastical bent of this latter branch of Unitarianism would support the Platonic conception of mathematics; in turn, the divine characterization of mathematics underwrote the central arguments of this nascent faith.

The connection between radical Unitarian thought and Platonic mathematics was particularly strong in the United States. Mathematical idealists and transcendental Unitarians joined in a revolutionary view of mankind, knowledge, and God that was highly antithetical to the tradition of Calvinism. Among their principles was the sense that humanity was not severely defective and thus could come to a high state of knowledge, that God has endowed humanity with the ability to comprehend the universe in a transcendental way, and that religion should tend toward the pure and ideal, not the liturgical or ecclesiastical. Each of these points was extended into the characterization of pure mathematics: through mathematics, mankind was capable of great things, arising out of a state of ignorance and coming into communion with God; God endowed human beings with pure mathematical concepts for just such a purpose; the best mathematics tended toward the pure and ideal, not the applied.

One American theologian who sided with the radicals during the Unitarian split and whose thought dovetailed nicely with mathematical idealism was Theodore Parker (1810–1860).[14] Parker's sermon "The Transient and Permanent in Christianity" (1841), for instance, forthrightly exhibited the religious idealism germane to Peirce and others who saw mathematics as rising above the base and particular opinions of human beings. "There seems to have been, ever since the time of [Christianity's] earthly founder," Parker declared, "two elements, the one transient, the other permanent. The one is the thought, the folly, the uncertain wisdom, the theological notions, the impiety of man; the other, the eternal truth of God. These two bear, perhaps, the same relation to each other that the phenomena of outward nature, such as sunshine and cloud, growth, decay, and reproduction, bear to the great laws of nature, which underlies and supports them all."[15]

Parker worried that the transient had overtaken the permanent in the

dogmatic faith of his age. He echoed his British counterpart Thomas Car-
lyle's clothing metaphor (from *Sartor Resartus*) when he wrote that specific
doctrines and liturgy "are the robe, not the angel, who may take another robe
quite as becoming and useful. . . . Looking behind or around us, we see that
the forms and rites of the Christians are quite as fluctuating as those of the
heathens."[16]

Appalled by the climate of sectarianism and ecclesiastical authoritarian-
ism, Parker lashed out at the established churches. He proclaimed that
"what passes for Christianity with popes and catechisms, with sects and
churches, in the first century or in the nineteenth century, prove transient
also."[17] The one true faith is instead that which communes with "the ideas
of Infinite God."[18] Parker believed that the age of dogmatic Christianity
would soon pass, giving way to a new harmonious era free of theological
disputes. The doctrine of the Trinity, the belief in the divine origin and absolute authority of the Old and New Testaments, and similar dogmas "are
fleeting as the leaves on the trees," he thought.[19] "It is hard to see why the
great truths of Christianity rest on the personal authority of Jesus, more
than the axioms of geometry rest on the personal authority of Euclid or
Archimedes," Parker noted. Rather, he continued, "the authority of Jesus,
as of all teachers, one would naturally think, must rest on the truth of his
words, and not their truth on his authority."[20] "Christianity," Parker famously concluded, "is not a system of doctrines, but rather a method of attaining oneness with God."[21] Such rhetoric energized idealist mathematicians, for it seemed to imply that pure mathematical research could be a
grand spiritual pursuit.

As the more conservative wing of Unitarianism increasingly emphasized
Scripture in the second quarter of the nineteenth century—clearly in opposition to theologians like Theodore Parker—Transcendentalism assumed
the mantle of idealism in America. The "Concord School of Philosophy" and
especially its high priest Ralph Waldo Emerson greatly influenced Benjamin
Peirce and other New England mathematicians. As Charles Sanders Peirce
jocularly recalled about his childhood, "I may mention, for the benefit of
those who are curious in studying mental biographies, that I was born and
reared in the neighborhood of Concord—I mean in Cambridge—at the
time when Emerson, Hedge, and their friends were disseminating the ideas
that they had caught from Schelling, and Schelling from Plotinus, from
Boehm, and from God knows what minds stricken with the monstrous mysticism of the East."[22] Many of the interactions between Benjamin Peirce and

Ralph Waldo Emerson took place in the Saturday Club, an association of intellectuals of all stripes in which Peirce and Emerson were charter members. The Club was full of writers, scientists, and clergymen who had been exposed to German philosophical idealism (generally through Coleridge) and who desired a new direction in American thought.[23] Emerson in particular guided Club discussions toward idealism. As he plainly stated, his philosophy of Transcendentalism was "Idealism; Idealism as it appears in 1842."[24] Peirce became a good friend and intellectual coconspirator of Emerson, eventually using the pillars of Emerson's philosophy to support a towering conception of mathematics. Indeed, Emerson's form of idealism would serve as the backdrop for the spiritual depiction of pure mathematics in nineteenth-century America.

Just as Benjamin Peirce was establishing his long tenure at Harvard University, the young Emerson was invited to give his groundbreaking address to the senior class of the Divinity School (1838). The religious vision Emerson evoked in that lecture, colored with Friedrich Schleiermacher's intuitionism and shaded by William Ellery Channing's liberal Unitarianism, displayed all of the marks of a maturing Transcendentalism. Emerson had in mind a broad audience for his address, not merely the graduating class of clergymen; he hoped his voice would carry beyond the walls of Divinity Hall into the greater intellectual circles of Boston and New England.[25]

Prefiguring Parker, Emerson described contemporary Christianity as a husk of its true self. Emphasizing the superficiality and hollowness of human opinions and dialects, he lamented that "the idioms of [Jesus's] language and the figures of his rhetoric have usurped the place of his truth; and churches are not built on his principles but on his tropes. Christianity became a Mythus, as the poetic teaching of Greece and Egypt, before."[26] The Bible, seeming like a fossil of ancient faith, cast spirituality in the past tense: "Men have come to speak of the revelation as somewhat long ago given and done, as if God were dead . . . the goodliest of institutions becomes an uncertain and inarticulate voice."[27] Furthermore, Emerson labeled the clergy (supposed stewards of religious belief and interpreters of Scripture) as hypocrites: they asked the poor to be hopeful, yet failed to invite them for supper; they demanded donations for missionary purposes, yet failed to spread faith locally. Organized religion "has lost its grasp on the affection of the good and the fear of the bad," Emerson bemoaned, "In the country, neighborhoods, half [of the] parishes are *signing off* . . . It is already beginning to indicate character and religion to withdraw from the religious meetings. I have heard a devout person, who prized the Sabbath, say in bitterness of heart, 'On Sundays, it seems wicked to go to church.'"[28] Perhaps most unsettling about the state of religious practice, Emerson thought, was the way in which it stifled the great intellectuals of society, who should be its greatest promoters. "Genius leaves the temple to haunt the senate or the market. Literature becomes frivolous. Science is cold" when true spirituality is absent from culture, Emerson told his audience.

Refraining from gloomy cynicism, however, Emerson proceeded to diagram a necessary Copernican reformation of faith, placing ideal elements at the center and moving human constructions to the periphery. He first condemned superficial ministers of religion: "The man who aims to speak as books enable, as synods use, as the fashion guides, and as interest commands, babbles. Let him hush."[29] He then sought to revive faith, to set it in motion once again: "The stationariness of religion; the assumption that the age of inspiration is past, that the Bible is closed; the fear of degrading the character of Jesus by representing him as a man; indicate with sufficient clearness the falseness of our theology. It is the office of a true teacher to show us that God is, not was; that He speaketh, not spake."[30] To revitalize religion, Emerson urged his generation to concentrate on connecting with the mind of God. "The sentiment of virtue is a reverence and delight in the presence of certain divine laws," he told the graduating class.[31] This communion meant that theology and pioneering intellectual work would be combined, with great minds leading the way: "I look for the new Teacher that shall follow so far those shining laws that he shall see them come full circle; shall see their rounding complete grace; shall see the world to be the mirror of the soul; shall see the identity of the law of gravitation with purity of heart; and shall show that the Ought, that Duty, is one thing with Science, with Beauty, and with Joy."[32] The Transcendentalist would be an active genius, reinvigorating faith.[33]

Moreover, the best scientists could be a part of this vanguard. Like Plato, whom he revered, Emerson delighted in the search for absolute laws that govern the universe in all of its particularity.[34] Presaging the odes to the discovery of Neptune, Emerson described grand scientific formulas as ideal, and thus spiritual: "Even in physics, the material is ever degraded before the spiritual. The astronomer, the geometer, rely on their irrefragable analysis, and disdain the results of observation. The sublime remark of Euler on his law of arches, 'This will be found contrary to all experience, yet is true;' had already transferred nature into the mind, and left matter like an outcast corpse."[35] Comparing the greatest work of the natural philosopher with the greatest work of the poet, Emerson lauded their comparable divine achievements: "It is, in both cases, that a spiritual life has been imparted to nature; that the solid seeming block of matter has been pervaded and dissolved by a thought; that this feeble human being has penetrated the vast masses of nature with an informing soul, and recognized itself in their harmony, that is, seized their law."[36] Science in its purest form comes into contact with the transcendental realm and transcribes its ideal concepts. Emerson thus performed an important service for mathematical idealists by placing the work of preeminent scientists on a par with the geniuses of poetry and art. Although he spoke little on mathematics per se, religious American mathematicians would be drawn to Emerson's thought, extending Transcendentalism into a field of which he was mostly ignorant.[37]

Emerson's notion that "the intellect" could rise above the biases of culture would also resound within mathematical circles. For Emerson, the intellect unveils ideas already extant on a higher plane, beyond the petty differences of human opinion. The abstraction of the mind permits humans to escape the prison of their age and culture: "The consideration of time and place, of you and me, of profit and hurt, tyrannize over most men's minds. Intellect separates the fact considered, from *you*, from all local and personal reference."[38] Constructing a view of transcendental objectivity that would become an important part of the nineteenth-century pure mathematician's self-image, Emerson emphasized the potent, detached nature of the intellect. "In the fog of good and evil affections it is hard for man to walk forward in a straight line. Intellect is void of affection and sees an object as it stands in the light of science, cool and disengaged. The intellect goes out of the individual, floats over its own personality, and regards it as a fact, and not as *I* and *mine*," he argued.[39] Mathematical reasoning, theorists such as Benjamin Peirce would assert following Emerson, is furthest removed from the profane and emotional realm of common humanity.

The Development of Peirce's Theology of Mathematics

Peirce's maternal uncle, Ichabod Nichols (1784–1859), was as important as Emerson and the transcendental Unitarians in shaping the ideas of the young mathematician. A pioneering liberal theologian who also happened to be a mathematician of no small talent, Nichols was the first to urge Peirce to come to a full appreciation of the spiritual significance of pure mathematics. He taught his nephew that mathematical work, especially in its most rarified incarnations, could be a religious testament. Educated at Harvard, from which he graduated at the age of eighteen as valedictorian, Nichols went on to become a mathematics tutor at the college. However, his stint as a professional mathematician was merely a way station to a highly desired career in the clergy. Studying theology along with mathematics, Nichols prepared himself to receive holy orders in the Congregational Church in 1809.

The ordination process was not what he or others expected. The ecclesiastical council, getting wind of Nichols's highly unorthodox theology, split in their decision to confirm him to a parish in Portland, Maine. His writing and sermons contained evidence that Nichols had drifted into a position that he and others called, in a somewhat vague and unsatisfactory way, "liberal Christianity."[40] Explicitly denying the doctrine of the Trinity and other central dogmas of traditional Christianity, he had in fact become a Unitarian. The other Congregationalist minister of Portland vigorously objected to this orientation, and Nichols and his supporters on the council—of which there were more than a few—seceded to form their own church. Freed from the doctrinal confines of the Congregational Church, Nichols finally had the latitude to preach his own brand of faith. That system of belief was radically ecumenical, with the mathematician-minister's sights constantly on highest common denominators. As one admirer commented, he tended to "the largest views" in all matters.[41]

Separation from the Congregational Church also allowed Nichols to declare that mathematics was just as important as the Bible, since mathematical conceptions, in his view, were thoughts from the mind of God. Nichols believed that the symbols, processes, and laws of mathematics did not originate in this profane realm; in their purity and efficacy they proved themselves to be emissaries from the heavenly realm. As a friend recalled, for Nichols "meditations on a mathematical law" were equivalent to "adoration, praise, or prayer" to the divine.[42] His shift from mathematics to the ministry was thus far from a tremendous change—the two careers were alternate vocations of pious devotion. Furthermore, because of his religious vision of mathematics Nichols maintained his endeavors in mathematical research well into his career as a clergyman in Portland. He analyzed and critiqued seminal tracts by Laplace, Cuvier, and Bowditch, and near the end of his life he followed the work of his nephew Benjamin with a keen eye.[43] He also wrote extensively on natural theology, contributing a volume on the subject that influenced other Unitarian theologians interested in science and mathematics.[44] The version of Unitarianism Nichols and his followers advocated was quite different, of course, from the variety that flowed from eighteenth-century British divines such as Richard Price. As with Parker and Emerson, Nichols put little emphasis on the rational interpretation of the Bible, or even much of an emphasis on the Bible at all. Instead, the desire to commune directly with the mind of God dominated his theology.

Nichols anticipated a fundamental position of nineteenth-century mathematical idealists when he advocated an ecumenical, unifying role for faith. In one early oration, he contrasted the still-convulsing nation of France with the newly formed United States. Taking a page from Edmund Burke, Nichols derided the Revolution as the triumph of "irreligion and licentiousness," as the victory of divisive men who had abandoned all that was good and true.[45] He feared that the rise of sharply divided political parties would lead to "the crude opinion, that, in politics, as in religion, there are as many great and good men upon one side, as upon the other."[46] This dangerous state required a reassessment of both politics and faith. Differences in opinion are largely due to human assumptions and the relativity of the senses, he noted, unlike mathematics, which transcends all points of view. Nichols lamented how clouded perception leads inexorably to disputes: "So various are the mediums, through which every thing may be viewed; and so greatly are the appearances of objects altered, according to the distances, at which they are contemplated; that differences of opinion with respect to them must necessarily arise; so that scarcely anything can be adduced, concerning which mankind universally agree."[47] Nichols sought to defuse this contentiousness by underlining that all of these differences ultimately dissolve in the mind of God. As one eulogy of Nichols recorded, "He disliked strife, stirred up no dividing questions, but carried, perhaps with over-caution, only the smoothing-plane of peace in his hand."[48] This yearning for reconciliation, in concert with Nichols's Unitarianism and metaphysical views of mathematics, would significantly shape Benjamin Peirce in his youth, providing him with animating questions and setting him on his way professionally and religiously.

Perhaps because of Ichabod Nichols's tutelage, a fascination with imaginary quantities seized Peirce as a child. These quantities, which include the square roots of negative numbers, "had a strong hold on his imagination," one friend recalled.[49] Charles Sanders Peirce recorded that by adulthood his father's "superstitious reverence for the 'square root of minus one'" had become an obsession.[50] Since no number exists that suffices as the answer to $\sqrt{-4}$ (recalling that negative numbers become positive when squared, dashing hopes for -2), "imaginary" or "impossible" were fitting labels for these numbers for most Victorians. But imaginary numbers seemed like wondrous, not capricious or maddening, entities for the young Peirce. These quantities magically resisted physical representation, yet the pure thought of higher mathematics could nevertheless prove they existed. The insight that certain mathematical objects cannot be put in a numerical table or graphed on a sheet of paper was an epiphany for a budding mathematician who would later proclaim a transcendental view of his discipline.

Later in life Peirce turned to imaginary numbers as proof of pure mathematics' revelatory power. In particular, he focused attention on quaternions, an algebra involving four variables and imaginary numbers that the Irish mathematician William Rowan Hamilton discovered and developed in the middle of the nineteenth century with no real application in mind. Subsequently, however, this abstract system became extremely useful for physicists who were trying to describe electromagnetic forces mathematically, such as James Clerk Maxwell. "The imaginary square root of algebra," Peirce rejoiced in his final work, "has become the simplest reality of Quaternions, which is the true algebra of space, and clearly elucidates some of the darkest intricacies of mechanical and physical philosophy."[51] W. E. Byerly recalled that his friend Peirce reveled in this prophetic quality of pure mathematics, the remarkable way in which "a calculus which so strongly appealed to the human mind by its intrinsic beauty and symmetry should prove to be especially adapted to the study of natural phenomena." "The mind of man and that of Nature's God," Byerly wrote about his friend's faith, "must work in the same channels."[52] Unsurprisingly, Peirce devoted much of his career to highlighting pure mathematical concepts that might eventually uncover scientific laws.

Peirce's transcendental view of his discipline became manifest in the early 1850s, when his fascination with pure mathematics combined with a growing religious liberalism. Despite the influence of Nichols, as a young man Peirce generally maintained that the Bible embodied a higher form of truth than science; when in doubt, science should bend to the Scriptures and not vice versa. As middle age approached and his understanding of the new geology of Charles Lyell and other scientists increased, however, Peirce's attachment to the Bible weakened. Evidence that the earth was far older than 6,000 years impelled him to adopt a position that would allow him to remain on the cutting edge of science while sustaining a deep faith.[53] Transcendental Unitarianism was perfectly suited for this symbiosis of science and religion, because (as we have seen) it held that all of the highest activities of the intellect unveil the divinity of the cosmos.

Like many of his idealist colleagues, Peirce built his case for the transcendence of mathematics beginning with geometry, because the field's power to transcribe heavenly laws had become obvious in astronomy (especially after the discovery of Neptune). In 1853, for instance, he illuminated geometry in the most flattering of lights in a prominent address to the American Association for the Advancement of Science. "Geometry," Peirce proclaimed, is rightfully "honored with the title of the Key of the Sciences . . . it is the key of an ever-open door, which refuses to be shut, and through which the whole world is crowding, to make free, in unrestrained license, with the precious treasures within."[54] Geometry lay at the heart of all other scientific endeavors—it was the most perfect way of describing the earth, the universe, and the movements of all bodies within. Borrowing heavily from Plato's *Timaeus,* Peirce extolled the geometric nature of cosmic design: "Try with me the precision of measure with which the universe has been meted out; observe how exactly all the parts are fitted to the whole and to each other."[55] Discoveries regarding the composition of the universe would only come, he asserted, through the application of pure geometrical forms.

Furthermore, according to Peirce the potency of geometry extended beyond the mere description and analysis of things we can see and touch. Advanced geometry communed with the sphere beyond matter—it was spiritual in nature. "Ascend with me above the dust, above the cloud, to the realms of the higher geometry, where the heavens are never obscured; where there is no impure vapor and no delusive or imperfect observation; where the new truths are already arisen, while they are yet dimly dawning upon the earth below," Peirce enjoined his audience with the fire of a preacher.[56] Indeed, the mathematician crescendoed to the kind of emotional religious peak now anathema in academic lectures: "Geometry! to the rescue! Geometry is at her post, faithful among the faithless. The pen is at work, the midnight oil consumed, the magic circles drawn by the wise men of the East, and the wizard logarithm summoned from the North."[57] In a telling use of words, Peirce forcefully argued that mathematics was like Cabala—or more accurately, it was the true Cabala that would instruct the modern world in the wonders and hidden secrets of creation:

Throughout nature, the omnipresent beautiful revealed an all-pervasive language spoken to the human mind, and to man's highest capacity of comprehension. By whom was it spoken? Whether by the gods of the ocean and the land, by the ruling divinities of the sun, moon, and stars, or by the nymphs of the forest and the dryads of the fountain, it was one speech, and its written cipher was cabalistic. The cabala were those of number, and even if they transcended the gematric skill of the Rabbi and the hieroglyphical learning of the priest of Osiris, they were, distinctly and unmistakably, expressions of thought, uttered to mind by mind; they were the solutions of mathematical problems of extraordinary complexity.[58]

Peirce thus agreed with Plato and Galileo that God wrote the universe in the language of mathematics, a divine dialect waiting to be discovered by the highest faculty of the human mind. Remarkably, no inscribed stone tablets were needed to understand this language—it could pass directly from the mind of God to the mind of man. In this way mathematics was uniquely suited to guide society to a higher level of understanding and a higher sense of faith. This grander religion would come, Peirce believed, because mathematics was not associated with mere liturgy or ecclesiastical structures; it was a pure language from the heavenly realm. "The loftiest conceptions of transcendental mathematics have been outwardly formed, in their complete expression and manifestation, in some region or other of the physical world . . . They are the reflections of the divine image of man's spirit from the clear surface of the eternal fountain of truth," he concluded with an idealistic flourish.[59] Mathematical symbols were clearly heavenly, rather than earthly, symbols.

Moreover, Peirce emphasized to the Association that some of the greatest minds of Western civilization sanctioned his lofty philosophy of mathematics. Particularly important to him were the Greeks: "Ancient philosophy, perceiving this power of number, did homage to it in all the simplicity, earnestness, and truthfulness which distinguished the early thinkers. Pythagoras and Plato, the founders of pure mathematics, turned their search inward to find in their own minds the origin of that force, which a universe of phenomena could only reveal in its effects, but not in its essence."[60] Like Plato, Peirce believed that the applicability of mathematics in day-to-day life was not its ultimate importance. Rather, the discipline's profoundly basic nature—its association with the Forms, to use Plato's terminology—justified and guaranteed its supreme status. There is "a broader basis than that of numerical accuracy" for awarding mathematics the highest place in the pantheon of intellectual pursuits, Peirce claimed, for mathematics was "of form; the grand type of structural combination . . . It is the alpha and omega of intelligible speech, the architect of the poetic temple, the founder of empires, and the maker of constitutions. It is the power of combining innumerable details into a consistent whole, the highest exertion of human genius, and that which approaches nearest to the act of creation."[61] In a sense, Peirce rephrased what generations of Christian theologians had said about the ancient Greeks—that somehow, deep in their hearts, they understood the transcendental faith that arose with Christ. "Modern science has realized some of the most fanciful of the Pythagorean and Platonic doctrines, and thereby justified the divinity of their spiritual instincts," he concluded.[62]

Mathematics came from heaven and helped its practitioners ascend to a higher plane, the mature Peirce thus believed. As he prefaced his most important technical work in mathematics, *Linear Associative Algebra*, "I presume that to the uninitiated the formulae will appear cold and cheerless; but let it be remembered that, like other mathematical formulae, they find their origin in the divine source of all geometry."[63] An exchange between Ichabod Nichols and his nephew in 1858, a year before Nichols's death, confirmed how Peirce had developed a robust mathematical faith. In a rapturous tone, Peirce wrote to his uncle of an equation that represented the sum of the mass, velocity, and force of the universe: "Th[is] Mathematical formula . . . which is quite unintelligible to the common reader, is one of the most extraordinary sentences which man was ever permitted to write. Rightly interpreted it is the concentration of the whole universe of material power . . . It is the indisputable proof that man has in his spirit the element of infinity. The full interpretation of this formula would exhaust the wisdom of men and angels."[64] Nichols joyously wrote back: "How delightful the fact that while we all enjoy a moral consciousness that our spirits contain the elements of infinity, that the mathematician is able to discover the same element in our souls through the medium of his own peculiar science! A noble illustration how the heart and the intellect conspire in support of the most important truths!"[65] In perhaps his most telling comparison, Peirce declared that mathematics was like the burning bush of the Hebrew Bible, in divine communication with the prophet, "ever burning and never consumed."[66]

Thomas Hill: Peirce's Bulldog

As Benjamin Peirce was formulating his views on religion and mathematics, he found constant encouragement from his lifelong friend Thomas Hill (1818–1891), a Unitarian clergyman who was president of Harvard during the Civil War. Hill's relationship to Peirce was similar to the relationship of T. H. Huxley to Charles Darwin—Hill was the most vocal advocate and popularizer of the mathematician's work. The two met when Hill was a sophomore at Harvard College and Peirce was a young faculty member.[67] Peirce enlisted Hill's help at the Harvard meteorological observatory, where they would spend late nights and early mornings recording data and talking about working in the service of God and the world through science.[68] Their chats led to a strong bond. Later, whenever Peirce wished to make sense of an inchoate mathematical insight, he would immediately call on Hill. One biographer recalled that in moments of discovery Peirce "would rush to the livery-stable behind the church, hire a chaise, and make all haste to Waltham where the Reverend Thomas Hill was then settled . . . Hill would gradually fathom the mind of Peirce and, towards morning, send him home to Cambridge with his problem stated on paper in his pocket and his thoughts at rest."[69] Hill was not, however, simply an intellectual midwife. Over the years he would help Benjamin Peirce construct a philosophy of mathematics that served as a pillar of their common faith.

Thomas Hill was another exemplar of the affinity between pure mathematics and idealist Unitarianism in the nineteenth century. His father was an English émigré who had fled Warwickshire in 1791 because he was religiously unorthodox and feared persecution; his mother, a Unitarian, prodded her husband toward her denomination from his more nebulous theism.[70] Thomas Hill Sr. was the typical father of a Victorian mathematical idealist: a tanner, he also dabbled in science and book collecting, and inculcated the love of nature in his children. His namesake son, in particular, acquired a scientific bent, which he took along to Harvard College in the 1830s. There, Hill excelled at mathematics and astronomy, inventing a mechanism (called the "occultator") for plotting the orbits of heavenly bodies, yet he felt his true calling was the clergy and eventually settled as a minister in Waltham, Massachusetts. Gaining prominence by writing essays on religion and education, Hill eventually received an appointment to the presidency of Antioch College in Ohio. After three years at that post, his alma mater called him to the Harvard presidency. At the time, the appointment caused a great deal of consternation among the more conservative members of the Harvard community, because Hill's theology was unusually liberal. (In the long history of Harvard, however, his presidency in the 1860s is only dimly remembered, overshadowed by his successor Charles Eliot's forty-year reign and by the darkness of the Civil War.) Criticism of Hill's administrative skills rather than his theology, however, prodded him to resign his post in 1868 and accept the pastorate at the same Unitarian church in Portland, Maine, where Ichabod Nichols had preached.

Surprisingly, Hill began his intellectual life as a materialist. Digesting the work of the chemist and philosopher Joseph Priestley and especially his "doctrine of Necessity," Erasmus Darwin's *Zoönomia* and *Botanic Garden,* and enough of John Locke (in particular, the *Essay Concerning Human Understanding* and *Essays on the Law of Nature*) to act as a catalyst generated an early belief in a familiar sort of Enlightenment materialism.[71] After the age of twenty, however, Hill's views radically shifted. Two seminal experiences changed his perspective, one aesthetic and one rational. First was an epiphany that nature was more special than the sum of its material parts:

> I remember distinctly . . . on the morning after a great June thunder shower my first vivid sense of landscape beauty. I looked westerly up a gentle sloping field which made the horizon only half a mile distant. The grass was of the liveliest green, while the road winding over it was of the peculiar deep red of the new red sandstone, or shale . . . that morning glance westward awaken[ed] in me what I recognized as a new feeling.[72]

He began reading widely in theology, learned Latin to gain ancient perspectives, and he eventually enrolled in divinity school. Reconsidering his early materialist philosophy another summer day while lying on the ground barefooted, he thought about the work of Euclid, which he had read as a young boy in an edition owned by his father. Pondering geometrical postulates, the empiricist philosophy of Locke suddenly seemed faulty. Not all knowledge was acquired through the senses, Hill now believed. This insight ultimately led him to "a system of logic based upon the assumption that certain truths are self-evident—seen by direct vision."[73]

Idealism now came to the fore. In the early 1840s he read books by the idealist George Berkeley, and engrossed himself in Charles Babbage's *Ninth Bridgewater Treatise,* which seemed for Hill the best statement of natural theology. He also began to examine works in pure mathematics and to associate increasingly with like-minded mathematicians, astronomers, and scientists.[74] Hill found the philosophy of the Harvard scientist Louis Agassiz particularly germane to his new mind-set, especially when Agassiz declared that "all facts proclaim aloud the One God, whom man may know, adore and love; and Natural History must . . . become the analysis of the thoughts of the Creator."[75] Agassiz's introduction to his *Essay on Classification*, among other scientific texts, gave Hill a strong sense of how the pure laws of science could come from God's mind.[76]

Around this time Hill also drifted into the orbit of Ralph Waldo Emerson. Hill had read Emerson's *Nature* upon its publication in 1836, with little initial impact (calling the Transcendentalist's work "queer stuff,—part sensible and part unintelligible"), yet by his second reading ten years later, the work had taken on new meaning for him.[77] During the year of Neptune's discovery, Hill read the entirety of *Nature* aloud five times, feeling "in profound accord with every line of it."[78] Emerson's other writings became similarly inspirational. Hill later recalled his deep assimilation of Emerson's poetry: "It was once my good-fortune to camp out on a mountain-top with a friend, who has since acquired a high name among astronomers. As we lay listening to the rain pattering on our canvas tent, and discussing many a theme not related to the stars, we found ourselves continually quoting from Emerson's first volume of poems. If the memory of either of us flagged, the other finished out the quotation; so that between us we had recited nearly the whole of the volume before we went to sleep."[79] Emerson clearly helped Hill move away from merely observational science and materialism to an overriding appreciation of the invisible realm of the divine. Years after reading Emerson's work, Hill summarized this influence when he recounted how Emerson had shown him "that every natural fact is a symbol of some spiritual fact."[80] The sensible world furnishes pointers to grander, heavenly notions, which in turn connect us to the sphere of God.

This transcendental philosophy of the generalized laws of science (written mathematically) also helped Hill erode the rigid boundary between Kant's phenomenal and noumenal realms. This erosion was evident, for instance, in Hill's reaction to the Scottish Kantian philosopher William Hamilton (1788–1856; not to be confused with the Irish mathematician William Rowan Hamilton). Hamilton had interpreted Kant's philosophy quite strictly—the finite was finite, the infinite was infinite, and never the twain shall meet.[81] Because human beings are finite creatures, we may never come to know attributes of the noumenal realm rationally. Hamilton called this basic inability of mankind to rise above the phenomenal plane "The Law of the Conditioned": "The conditioned is that which is alone conceivable or cogitable; the unconditioned that which is inconceivable or uncogitable . . . In other words, of the absolute and of the infinite we have no conception at all."[82] For Hamilton, therefore, knowledge of the divine Mind was beyond the capacity of human beings. We may—and should, since Hamilton was a devout Christian—believe fervently in God, yet His infinitude entails that we cannot understand much about Him beyond a simple intuition of His existence.

Thomas Hill conceded that in metaphysical legalese that which is infinite is ultimately incomprehensible, yet to him advances in pure mathematics suggested a more sanguine view of the potency of the human mind. Modern geometry and calculus, for example, handle entities that are infinite, such as infinitesimals and infinitely long lines. Integrals exhibit the remarkable power of mathematics: "The metaphysician says that the march through indefinites can never reach the infinite. But that is an error. The march through indefinites can reach the infinite, provided the march be always at an accelerating pace."[83] By adding together finite quantities using advanced mathematical processes, we can thus approach the infinite: "Although we cannot conceive the infinite, as such, we can conceive, and conceive correctly, the result of this attainment of the infinite, when the result is finite."[84] In pure mathematics human beings grasp the unconditioned and break Hamilton's Law of the Conditioned. Like the German theologian Friedrich Schleiermacher, Hill thought we could access the infinite, yet unlike Schleiermacher he believed this was possible through reason as well as intuition. Pure mathematics highlighted a better way—a method of communing with the divine realm by using rationality rather than feeling.

Hill thus argued from his knowledge of mathematics that Hamilton (and his English disciple Henry Longueville Mansel) wrongly characterized the human mind as hopelessly divided into two irreconcilable faculties, and in doing so gave support to the worst materialistic tendencies found in the philosophies of prominent contemporaries such as John Stuart Mill and Herbert Spencer:

Kant's distinction between the pure and the practical reason, Hamilton's between the cognitive faculties and faith, Mansel's between speculative and regulative truths, are all untenable. The two sets of our faculties and the two sets of truths, thus distinguished, are substantially one, and their separation is an uncalled for concession to that school of philosophers who would bound our knowledge by that which can be logically deduced from the testimony of the senses . . . if we show that this definition cannot account for the action of the human mind, nor explain the triumphs of either ancient or modern geometers; we may also resist Mill's definition of the mind as a congeries of the possibilities of sensation, and Spencer's as the state of consciousness, and Spinoza's as the sum of our thoughts; show that such definitions cramp and pervert both psychology and ontology; and refuse to make the smallest concession to any philosophy that would make mind a mere modification of matter.[85]

Hill declared that materialists and phenomenalists who believed the human mind could never transcend its own senses and ideas, as well as those who thought that only a faculty above reason could make such a leap into the divine realm, were all grossly mistaken. Human beings may be limited, yet we are graced with the magnificent and quite rational method of pure mathematics, which is able to comprehend the noumenal realm without recourse to some other part of the mind or soul. Mathematical reasoning, and by extension human understanding in general, therefore has a grand capacity to know the realm of the Absolute—the Kantian barrier between two halves of the mind was illusory. Like God's mind, the human mind was not divided. In mathematics speculative reason metamorphoses into transcendental reason, its pure wings allowing the human mind to float to the heavenly plane.

Although Hill's attack on William Hamilton contained many abstruse philosophical arguments, his thoughts on the interrelation between mathematics and religion initially surfaced in a work intended for a general audience, *Geometry and Faith* (1849). Written immediately following the discovery of Neptune, Hill's opening statement that "observation alone can lead to nothing, without insight—without that clearness of inward vision which sees more than the outward fact, sees the divine ideal which the fact partially embodies" shows how the armchair triumph of the pure mathematicians Le Verrier and Adams was fresh in his mind.[86] The book was a major success, going through three editions and doubling in size each time.

Like so many other works of mathematical idealism in the nineteenth century, *Geometry and Faith* began with epigraphs from the philosophers highlighted in chapter 1: quotations from Plato's *Epinomis* and the seventh book of the *Republic,* and from the work of the English Neoplatonist John Dee.[87] Hill traced the process whereby mathematics comes into contact with the ideal, divine patterns of nature and thus gives strength to faith. The process begins simply, he noted, even in childhood. We enjoy symmetry, and this "perception of beauty in outline is the unconscious perception of geometric law."[88] This innate love of symmetry is the first step to both mathematical understanding and the comprehension of divine laws, Hill then explained. It shows that we are truly the earthly creatures closest to God: "The human mind, fettered by the body, seems in such speculations to show its kindred to the Infinite Spirit."[89] Human beings delight in the discovery of mathematical regularity, which raises our sights and purifies our thoughts. We begin to understand how our mind resonates with patterns that emanate from the Source of all cosmic patterns. In his concluding articulation of this thesis, Hill wrote, "Mathematical science cannot admit the possibility that the rhythm and symmetry of the organic kingdoms is an accidental result of accidental variations; there must be algebraic and geometric law at the basis, not only of each organic form, but of the series of forms. The series has a unity; capable, when men have attained a fuller comprehension of it, of expression in terms of thought."[90] Comprehension of nature's mathematical regularity lifts our mind up into association with the heavenly realm. The inspiration of a Leibniz could be as beautiful and pious as that of a Handel.[91]

Educational Reform and Elitism

Understanding mathematics as a divine language produced in Thomas Hill and Benjamin Peirce an impulse to reform higher education. Fighting against rote memorization and textbook learning, they sought a more flexible educational system designed to unleash genius and reveal eternal laws. These pedagogical theories clearly paralleled their religious idealism—true knowledge, they believed, came not through texts (Scripture) and lectures (sermons) but through a direct engagement with the highest notions the mind is able to contemplate. At a time when American universities were just beginning a radical shift from parochial training grounds for the clergy to research centers for a number of natural and social sciences, the two mathematical idealists proposed a new agenda for Harvard.[92] Hill's and Peirce's educational concepts bridged the past and future of academia; they hoped to increase support for their scientific endeavors while creating a new clerisy and renewing faith.

First on the agenda, unsurprisingly, was raising the status of mathematics in higher education. As Hill told a Harvard audience in 1858, "There can be, in my view, no true education that is not founded upon a knowledge of the mathematics."[93] In his extended essay on pedagogy, *The True Order of Studies* (1853), Hill had made this point even more forcefully: "The mathematics take logical precedence, as the great and indispensable foundation of all learning. It is . . . impossible to place them anywhere else than at the beginning of all intellectual education. All intellectual life upon our planet begins with geometry."[94] Even the most basic mathematical texts should be instilled with a divine reverence for the subject, Hill thought. For instance, in his mathematics primer *First Lessons in Geometry* (1855), he began not with the definitions of points, lines, and planes, but with a short homily on geometry's supreme meaning and utility. Simplifying his language to the level of his elementary audience, Hill spoke of the wondrous nature of his topic: "If you want to understand about plants and animals, and the wonderful way in which the All-wise Creator has made them, you must learn a little Geometry, for that explains the shapes of all things."[95] Following his introduction to the fundamental elements of geometry, Hill returned to grander, more spiritual themes. He entreated students to pursue the discipline to higher levels of pure thought: "I hope that at some time you will study more Geometry, and learn how to prove the truth of all I have told you."[96] Hill concluded the primer by invoking the bond between God and man. "The great Creator, who has made all things in number, weight and measure, knows everything," he wrote, "and the more we know, the more clearly we shall see how great is His knowledge, how wonderful his wisdom, and how beautiful the manner in which he has used what we call Geometry in the forms he has given to all things on the earth or in the sky."[97] In the hands of Thomas Hill, teaching mathematics could be another way—or even the primary way—to inculcate spirituality in the next generation.

Indeed, responding to the growing movement to secularize public education, Hill used mathematics as a rhetorical Trojan horse. Instead of directly addressing proposals to rid education of religious elements, he first asked whether we should teach geometry in public schools. The answer to that question, of course, was a resounding yes. But what do we actually mean by geometry? A course in geometry begins by looking at simple shapes, Hill continued, but then moves inexorably toward the abstract: "Afterward the pupil should be lifted up to a scientific and even to a meta-physical view."[98] Like Plato, he traced a line from the particulars of the sensible world to the divine forms of pure mathematical knowledge. His idealism now bare, Hill attacked the premise that religion stultified education and progress. The spiritual view of mathematics will benefit human knowledge and society, he claimed, for the intense desire to comprehend the divine Mind motivates new research and produces new revelations. "A large proportion" of scientific advances, he underlined, "have been made by those who were stimulated to the solution of these problems by the faith that they thereby came into communion with the thoughts of the Most High."[99] In other words, divorcing religion from education would condemn America to intellectual stagnation.

As the president of Harvard, Thomas Hill fought to reorient the institution toward advanced research. His annual reports to the Board of Overseers pleaded with increasing passion for a university that would promote the highest achievements of humanity. Plotting a course for the university at the beginning of 1864, Hill wrote, "It is time that we should have in our country at least one institution thoroughly organized and amply endowed, at which it shall be a principal aim to carry those students who have the highest talents to the highest degree of culture; and also through its teachers, its pupils and graduates, to extend the domain of science, and increase the fruits of learning."[100] Somewhat preposterously, Hill even hinted that the ongoing Civil War was perhaps traceable to this lack of an indigenous research institution.[101] In his mind one of the main problems was that pedagogically Harvard emphasized base repetition rather than transcendental abstraction. Expanding on his redefinition of the university, he lamented that "if the genius of our Colleges is such that they must be confined to the diffusion of knowledge" the institution would not be able to pursue a properly grand vision of philosophy, art, and science.[102]

Hill claimed that his educational changes would free up geniuses like Louis Agassiz and Benjamin Peirce to engage the mind of God. "Our best Professors are so much confined with the onerous duties of teaching and preparing lectures, that they have no time nor strength for private study and the advancement of science and learning," he complained.[103] The nature of whatever teaching remained would also have to be altered, Hill argued, with less time spent on applied knowledge and rudimentary facts and more time on loftier concepts that elevate the student's mind to a higher plane. This shift was particularly necessary, he thought, in mathematical education. "In the mathematics the candidate . . . misuses time thro the inverted method of teaching so common in schools; exercising the memory, loading it with details, straining the reason with demonstration; but not illuminating the imagination with principles to guide its flights," Hill bemoaned.[104]

Peirce followed suit, welcoming (one suspects with a modicum of self-interest in addition to reformist zeal) the prospect of less low-level teaching and more freedom for himself and others to unveil the great laws of the universe. He especially clamored for an elective system at Harvard that would allow sharp students to advance quickly in a field while letting others drop a subject (like mathematics) after their freshman year. Some have seen this elective system (which Hill's more effective successor, Charles Eliot, truly implemented) as the beginning of a more practical course of study in American colleges, allowing educational institutions to justify themselves to their surrounding communities.[105] Of course, the elective system at Harvard was also doubly beneficial for certain faculty. It removed poorly performing students from the classrooms of difficult subjects (such as advanced mathematics) on their own volition, while also removing professors of such subjects (like Peirce) from classrooms in general, giving them more time for their own research.

Peirce believed that many of his colleagues would benefit from this rebalancing of their workload. "I cannot believe it to be injudicious to reduce the time which the instructor is to devote to his formal teaching, to a couple of hours each day, or even to less than this, and to much less, if the same man is to undertake more than one branch of study," he wrote in the *Harvard Register*.[106] Peirce failed to understand why all students could not learn the way he had—through the solitary study of a discipline's groundbreaking treatises. "Enthusiasm, which is the highest element of successful instruction," Peirce claimed, "can best be imparted nearest the fountain-head, where the springs of knowledge flow purest, and where the waters are undiluted by the weakening influence of text-book literature." Therefore Harvard should work to promote the "diligent study of original memoirs," he thought, rather than force students to attend dull lectures on fundamentals.[107]

Peirce's haughty classroom demeanor vividly displayed his theory of education. Harvard undergraduates continually protested that they had trouble following his rapid-fire lectures. As one student groused, "He had little respect for pupils who had not a genius for mathematics, and paid little respect to them."[108] Early in his career, undergraduates found the "basic" textbooks Peirce authored so opaque they complained to the administration. After some deliberation a committee reported that Peirce's writing, while in a certain sense "elegant," was so "abstract and difficult, that few could comprehend . . . without much explanation." They ultimately concurred with the students that "books for young students should be more simple."[109] Peirce responded to these complaints about his pedagogy in a characteristic fashion—he stopped bothering with introductory courses entirely, deigning to teach only at the uppermost levels of his discipline. And he did not care who his students were, as long as they were transcendentally brilliant: "I will do the same for the young women that I do for the young men. I shall take pleasure in giving gratuitous instruction to any person whom I find competent to receive it. I give no elementary instruction, but only in the higher mathematics."[110] Peirce's choice of students clearly matched his elite conception of mathematics.

Despite early complaints many found Benjamin Peirce's dense teaching style increasingly magnetic. Charles Eliot, who began his career as a chemist, remembered Peirce as "a very inspiring and stimulating teacher. He dealt with great subjects and pursued abstract themes before his students in a way they could not grasp or follow, but nevertheless filled them with admiration and reverence. His example was much more than his word."[111] Indeed, the Harvard administration gave Peirce remarkable latitude later in his career, rather than asking him to simplify his instruction or to make himself more approachable.

Peirce's elitism apparently added aloofness to the job description of the eminent academic at precisely the moment when American universities like Harvard were attempting to bootstrap themselves into world-class prominence. The model of Professor Benjamin Peirce—an awe-inspiring genius rather than a fellow human being—became quite attractive in this context. Such a characterization benefited several factions: professors, who enjoyed a great deal more authority in that mold; students, who had previously considered Europe the only place one could encounter leading scholars (and who were paying sharply more each year in America and thus felt they deserved the presence of genius); and cities and states, which could rest assured that their significant financial contributions to universities were worthwhile.

Tall tales and witty retorts helped to mythologize Peirce as an unmatched intellect. "To be great is to be misunderstood," Emerson famously wrote; the Harvard community created its own permutation of that Romantic formula in figures like Benjamin Peirce: To be great is to be incomprehensible. For instance, when Peirce began to suspect that Neptune's discovery was somewhat fortuitous, in part a stroke of luck rather than the true product of unadulterated mathematical genius, his urge to debunk the planetary discovery had to be restrained by Edward Everett. "Nothing could be more improbable than such a coincidence," Everett admonished Peirce, referring to the minuscule chance of a telescope randomly focusing on the appropriate part of the sky where the dim planet hid. "But," Peirce snapped back, "it would be still more strange if there were an error in my calculations."[112] A rumor abounded at Harvard that Peirce and his four sons, all of whom had superior mathematical talent, never finished a hand of cards—they merely calculated the statistical probability of the winner and then paid out bets accordingly.[113] Even in front of fellow professional mathematicians Peirce flaunted his elitism. He once spent an uninterrupted hour writing thick equations on a blackboard at a meeting of the National Academy of Sciences, after which he finally turned around to his audience. "There is only one member of the Academy who can understand my work," Peirce reportedly sighed to the perplexed crowd, "and he is in South America."[114] Similarly, when asked by a foreign counterpart what American mathematicians thought of a recent European appointment, Peirce declared that it was a good thing the gentleman had asked him, because he was virtually the only American knowledgeable enough to provide such an opinion.[115] Such stories, embellished by countless retellings, created an aura around Peirce, and he did little to dissipate it.

Being thought of as a transcendent genius truly had its practical benefits—it lent Peirce's declarations a great deal of weight. As one obituary wryly noted, "Persons who could not understand a word of his abstruse speculations were compelled to listen to his earnest argument, and knew that his conclusions must be important and true, even when they did not know what his conclusions were."[116] Peirce appeared to audiences not unlike a revivalist preacher speaking in tongues. Like Adams and Le Verrier, he was called with increasing reverence a "real live genius," a "necromancer," a "wizard," a "prophet," the "warder of an enchanted tower . . . whose speech was foreign."[117] His mysterious and condescending stage presence fostered an immense amount of respect for his proclamations, a reverence Peirce would find valuable as he became a very public intellectual.

The Relationship of Science and Religion

Conceiving mathematics as an endeavor potently capable of revelation led Benjamin Peirce to see few parameters for his discipline. Indeed, as he stated in the philosophical introduction to his *Linear Associative Algebra,* mathematics "belongs to every enquiry, moral as well as physical."[118] As the years passed, Peirce increasingly branched out from his work in pure mathematics and the natural sciences to offer the public guidance on weighty matters of the day. One of his main responsibilities, he believed, was to quell the brewing antagonism between what he considered to be scientific and religious extremists.

Even before the publication of Darwin's *Origin of Species,* Peirce had declared any such contretemps manufactured and foolish. In his 1853 speech to the American Academy of Arts and Sciences, he decried "men, and pious men too, who seem honestly to think that science and religion are naturally opposed to each other . . . I cannot conceive a more monstrous absurdity."[119] Referring to the topic of his lecture, geometry, Peirce delineated a familiar argument he would use time and again to rebut those who placed religion and science in diametrical opposition: "There is proof enough furnished by every science, but by none more than by geometry, that the world to which we have been allotted is peculiarly adapted to our minds, and admirably fitted to promote our intellectual progress. There can be no reasonable doubt that it was part of the Creator's plan."[120] In other words, the congruence between pure mathematical concepts and the actual universe cannot be a mere coincidence; its perfection indicates a divine origin. Likewise, when he encountered criticisms of science by strict adherents of the Bible, Peirce responded vociferously. "How can there be a more faithless species of infidelity than to believe that the Deity has written his word upon the material universe and a contradiction of it in the Gospel?" he asked.[121] One God entails one Truth—to believe otherwise was an affront to theology, science, and common sense.

Later in life, Peirce extended his mediation of the supposed conflict between science and religion beyond mathematics. Of course, after 1859 the most contentious issue became biological evolution, and Peirce was not shy about broadcasting his views on the subject. The mathematician thought a reasonable middle ground clearly existed between biblical literalists who depicted Darwin as Satan's ambassador and radical scientists who thought modern biology obviated the need for grossly inaccurate ancient texts.[122] In

1877, for example, Peirce conveyed this moderate position to an audience at the First Church in Boston after a flare-up over Darwin's theory. His language, normally complex, became quite plain as he emphasized that all human opinion dissolves in the mind of God: "Science and religion are born of the same house, and that house is not divided against itself. There will at all times be an apparent conflict between them arising from defects of human nature; but all this conflict is of human origin, and it originates in the deficiency of our knowledge, not in the greatness of it."[123] Criticizing the defects of human understanding shrewdly helped to reconcile interpretations of the Bible with the discoveries of contemporary science.[124]

In addition, like many others, Peirce could not accept that evolution was simply material change without greater significance. In a lecture given two years after his First Church oration, he cautioned, "God created the earth, and endowed it with its exhaustless power of development. Shall we, scientists of to-day, erect this created evolution into an original divine power, and honor it as a god? We might as well go back to the worship of light and the sun."[125] Only one conclusion was possible, one that kept God at the center of all things: "The Creator is not ruled out of the universe by our theory of evolution. That which we call evolution is but the mode . . . [and] manifestation of his paternity. He becomes through it, more legibly than ever, the beginning and the end, the Alpha and the Omega, the eternal I am."[126] "Judge the tree by its fruit," Peirce implored his audience; either the universe is "a tale told by an idiot, signifying nothing" or it is "the poem of an infinite imagination, signifying immortality."[127] The faithful can accept evolution when they understand it not as a course of random, pointless mutations, but rather as a divine process with a guiding Hand.

More broadly, the religious idealism of Unitarians like Benjamin Peirce made them impervious to any clash between religious orthodoxy and scientific progress. After all, to such believers any new scientific discovery was merely another marvelous expression of God's thoughts. As one likeminded Unitarian minister argued in *The Monthly Religious Magazine* following an imbroglio over evolution, "Science is destructive of the prevalent Christianity, which is slavery to the letter and form, but is constructive of the spiritual Christianity which Jesus proclaimed. Christianity and Science, married by Reason, are ordained of God to produce a nobler religion than men have yet known."[128] This "nobler religion" would leave behind the particularity of the material world and the opinions of humanity in an upward movement toward the purely ideal. Just as "the perfection of theology requires that all the gods should be reduced to one God," Peirce wrote in his final work, "all science can be reduced to one fact. Among the facts to be embodied, are the facts of omnipresent ideality, the intelligible cosmos, and the all-comprehending intellect."[129] In short, theology and science both lead to a single Source. Peirce found it heartening that his brand of Unitarianism could count among its ranks numerous leading scientists, especially in and around Boston. Their robust faith would provide a light, he believed, to those who could not envision a modern science friendly to the cause of religion.

Even more aggressively than his friend Benjamin Peirce, Thomas Hill saw himself as a crusader against those who fostered unbelief in the name of science. Hill found Darwin's evolution by natural selection more troubling than Peirce, seeing the biological theory as a symptom of a widespread ideological threat that had grown out of the thought of the nineteenth-century French philosopher Auguste Comte. Building on Comte's *Cours de Philosophie Positive* (1830–1842), the Positivists held scientific data and theory aloft as the totality of knowledge about the universe, or in Hill's caustic words, "the doctrine that we cannot go behind the generalization of facts in the outward world."[130] Secular science had finally achieved its proper, regal place in society, the Comteans thought, surmounting first theology and then metaphysics as history and human understanding marched inevitably forward. Pointing to a "heavenly" or "ideal" realm for our answers was immature—everything we needed was right in front of our eyes.[131] Like many other English-speaking intellectuals, Hill apparently discovered Comte's ideas through John Stuart Mill's *Logic* (1843), which he quickly rebuffed in a review for the *Christian Examiner*.[132] Harriet Martineau's 1853 English translation (and vast abridgment) of *Cours de Philosophie Positive* gave Positivism a wider audience—and among many theologians, a harsher reception.[133]

As Hill saw it, the Positivists had trouble distinguishing the forest from the trees. Comtean philosophy was "equivalent to saying that there is no meaning in the book of nature, that the grammar of her language is all that is to be studied; that we may analyze and parse her sentences, but not discover their import."[134] In sermons, Hill reserved his most vitriolic criticism for those who had adopted this philosophy, for "the speculations of a school which professes to have outgrown Christian faith in God."[135] Although he understood that few Americans had actually read the works of the French thinker, Hill believed that many had internalized Comte's ideas unwittingly through secondary channels. Comte's outlandish effort later in his life to construct a new "Religion of Humanity" to supplant Christianity might have tarnished his name; nevertheless, Hill felt the philosopher's early work still "exerted a great influence; and some of the clearest English and French writers of our day owe to him, indirectly, more than they, perhaps are, themselves aware."[136] For example, Hill singled out the social scientist Henry C. Carey for his Positivist description of geometry as merely measurement without transcendental significance.[137]

Hill advanced a version of William Paley's "argument from design" to counter those swept up by Positivism. For example, in one sermon entitled "God Known by His Works" (1887), Hill highlighted natural theology as a basic element of human nature: "No man who inquired the purpose of a piece of cast metal would be satisfied on being told that it came into its present shape by being run in a mold. He would ask the purpose for which the mold was formed."[138] Man's curious nature leads inexorably to the higher questions of religion, and "if he pushes on . . . he is led to the will of God."[139] Hill did not bother listing nature's magnificent elements in "God Known by His Works"; by the time he delivered that sermon he had already published a major, detailed work on the subject, *The Natural Sources of Theology* (1877), comprising numerous articles written for *Bibliotheca Sacra*.[140] It merely sufficed for an audience familiar with the argument from design to note that "we cannot turn the eye in any direction without beholding ten thousand thousand proofs of the wisdom and of the love of the Creator."[141] In Hill's benign (and sanitary) description of nature, the design of the seashell, the miraculous growth of the forests, and the regularity of the tides all pointed to a higher Being.

Thomas Hill was not at heart a Paleyite natural theologian, however, spending his time compiling every last bit of evidence like a defense lawyer for faith. By Hill's own reckoning it was more accurate to say that his argument was "morphological" than "teleological." That is, he was less interested in the *purpose* of the many details of nature than in the general *shape* of those elements—their mathematical construction, their geometry. The "morphological argument" entailed that "when anything whatever is found to be so arranged as to express or embody an idea, the presumption is that the arrangement was made by an intelligent will; and this presumption increases in strength with the complication of the arrangement, the complexity of the idea, and the fidelity of the arrangement to the idea; increases with such rapidity that a very moderate degree of complexity and of fidelity

makes the presumption become a certainty."[142] In other words, the congruence of pure mathematical concepts with natural artifacts proved for Hill
the divinity of both the cosmos and mathematics. For example, Hill noted
that Benjamin Peirce's 1843 work on integral calculus included a mathematical discussion that prophesied exactly a physical phenomenon witnessed by scientists in the 1860s.[143] Similarly, the ancient Greeks' purely abstract description of the ellipse was astonishing since the form proved to be
the true shape of the planetary orbits unveiled by Kepler two millennia
later.[144] "In all these cases of the embodiment in nature of an idea which
men have developed, not by a study of the embodiment, but by an *a priori*
speculation, there seems to us demonstrative evidence that man is made in
the image of his Creator; that the thoughts and knowledge of God contain
and embrace all possible *a priori* speculations of men," Hill proclaimed.[145]

The clergyman and educator therefore made his stand against the threat
of Positivists and other secularizers not by pointing to the miraculous construction of the human eye or ear, but by highlighting the remarkable constitution of the human mind and pure mathematics. As Hill declared in
1860 to the graduating class at Antioch College, "The mathematics alone
give us a knowledge of the exactness of God's thought, and alone are capable of demonstrating to us, from the manifestation of thought in the creation, the necessity of supposing the existence of a Thinker."[146] The mind,
Hill asserted, is born with the ability to summon abstract concepts from the
ideal realm, which in turn help us to comprehend the Divine. "The true aim
of science," he emphasized, "is not merely to record the uniformities of nature, but to discover the intellectual ideal which binds them together."[147]
Furthermore, Hill believed all human beings share in the revelations of the
highest work in mathematics and science. Borrowing phrases from the
Passover Haggadah, in which each generation of Jews repeat that *they* were
the ones that God spirited out of Egypt, he wrote that it was "I that pointed
Galle's telescope to the invisible planet, and that climbed the rays of light,
far beyond the solar sphere."[148]

Hill (somewhat hyperbolically) cautioned his generation, moreover, that
cultures that recognize the transcendence of mathematics and its implications achieve greatness, while those that do not engender social degenera-
tion. He contrasted the legacies of the golden age of Greece and the Roman
Empire. "At Athens, men were very early taught and very tenaciously held
to the truth, that geometry deals neither with perishable matter, nor with abstract measurement, but with the eternal verities of space, the uncreated

ideal forms, on which the Infinite Mind patterned, from the beginning, the things which appear," Hill approvingly noted.[149] The brilliant Greek culture held (as works such as Plato's *Epinomis* and *Republic* attest) that "he who studies geometry is holding communion with the divine, the everlasting Thought, which formed and guides the universe."[150] Ancient Greek intellectuals thus had the proper stimulation and motivation necessary to make significant advances in all realms of knowledge.

The Romans, on the other hand, had no larger, spiritual conception of mathematics, Hill thought, and thus became mired in artistic and scientific mediocrity, amorality, and apathy.[151] By envisioning mathematics as merely a useful tool rather than a transcendental science, the Romans started down a path that ended with a denial of all that was lofty and stabilizing in human culture. The debasement of mathematics led inevitably to a decline in ethical standards and the potential for intellectual advancement. Hill's invective against the Romans undoubtedly went too far; without qualification he claimed (Gibbon be damned) that the "ruinous heresy of the Latin race, that of regarding geometry as the mere science of measurement" led directly to the decline and fall of their empire.[152] "The Roman view of geometry," he sternly cautioned his audience, "threw the long shadow of the dark ages over the whole of Europe."[153] Regardless of the historical accuracy of this pronouncement, the implication was clear: If we want to avoid another thousand years of darkness, we must beware of secular views of mathematics.

The "Ideality" of Mathematics

In concert with Thomas Hill, Benjamin Peirce proposed a special word believers could use to indicate common support for a transcendental philosophy of mathematics, "ideality." John Norris, the early modern Cambridge Neoplatonist examined in chapter 1, originally coined the term to denote when pure thought accesses the divine plane.[154] Peirce appropriated the word for the title of his final work, *Ideality in the Physical Sciences,* published posthumously in 1881. He had been working on *Ideality* for most of the last years of his life, reading portions of the text to his most treasured associates for approval. He had also delivered much of *Ideality* as a set of lectures at the Lowell Institute in Boston and the Peabody Institute in Baltimore in the late 1870s. Peirce intended the book to be a comprehensive and readable summary of his scientific and theological beliefs, or more accurately to be a proper and full delineation of how the two were in fact one and the same.

Laying bare his religious idealism for the general public, Peirce emphasized in *Ideality in the Physical Sciences* the divine nature and construction of human reason. The mind was not a passive receptor and organizer of sensations. Rather, it contained a priori concepts—as Kant had identified—that were instrumental in understanding the universe. Like Thomas Hill, Peirce then blurred the line between Kant's distinct realms and mental faculties, arguing that the pure concepts and laws upon which all great science is based are part and parcel of the divine Mind. "The existence of intimate relations between the constitution of man's mind and that of God's firmament" show that we are blessed creatures, he rejoiced.[155] Reason therefore could have a spiritual aspect—the resonance between advanced rational conceptions and the divine composition of the universe. Newton may have discovered the law of universal gravitation, for instance, yet it did not originate with him nor belong to him. His equation was "a sublime embodiment of all-embracing thought," a transcript of a divine Idea invoked by God when He created the cosmos.[156]

Peirce believed the mind achieved its highest level of purity, and thus divinity, in mathematical equations and concepts such as those found in Newton's *Principia*. Although he mentioned poetry and the arts as spiritual endeavors, they paled in comparison with the transcendental discipline of mathematics. "Ideality," Peirce declared, "is pre-eminently the foundation of the mathematics."[157] Never having to resort to the material realm at all, pure mathematics surpassed other revered elements of human culture: "In the studio of the painter, the sculptor, and the poet, ideal art is prone to conceal its natural figure under the garb of reality. But in the frozen cave of geometry, the thoughts which may trickle in from the actual world are crystallized . . . they are ideal truth."[158] The imaginary quantities and quaternions he enjoyed in his youth were still vivid for Peirce as exemplars of the pure nature of advanced mathematics. In a poetic ode to such ideal mathematics, visible only in the mind's eye, Peirce wrote, "It is a daring flight upon the wings of pure logic by which the modern geometer has soared away from the realities of the ancients into the strange ideality of the fourth dimension of space, with its impossible triangles and its inconceivable metamorphoses."[159] Although the data collection of everyday science was important to the progress of knowledge, Peirce emphasized that true geniuses "have penetrated *through* fact to the inmost soul of Nature; and their proudest discoveries have invariably been vast intellectual conceptions exhumed from the recesses of the material world."[160] In the ideal form of pure mathematical notation human knowledge achieved its most glorious congruence with the mind of God.

Peirce again followed Hill's cue by claiming that although geniuses unveil grand mathematical laws, all human beings are naturally predisposed to understand them. Pure mathematics, he underscored in *Ideality in the Physical Sciences,* "vibrates in every soul,"[161] a key part of God's plan to give every individual the capacity to know of His existence and intelligence. In addition, mathematics provides a common platform for human communication: "Certain general principles of human nature must be diffused through all souls which are intended for mutual intercourse. Such is mathematical truth, which all men accept as absolute."[162] Like Theodore Parker, who pointed to a single "true" spirituality within all denominations, Peirce believed that in the ideal sphere of pure mathematics all sectarian differences disappeared. Indeed, Peirce's only nod in *Ideality in the Physical Sciences* to the Judeo-Christian tradition was his offhand remark that the divine Mind of which he spoke was also "the great lawgiver of Judea."[163]

With Christ almost absent from the text, more orthodox Christians must have found Peirce's *Ideality* more Judeo than Christian; in fact, in his final work the mathematician most often comes across as a broadly construed scientific monotheist. Summarizing his theology, Peirce told of how pure science would fulfill Emerson's call for a reinvigoration of faith: "Science will recover the perception of the central luminary, which is the unfailing fountain of pure knowledge, and will be restored to the praise and worship of the almighty, omniscient, and all-loving God."[164] Benjamin Peirce, enraptured by the ideal concepts of mathematics, had little need for the intermediary of Christ. God would be revealed through equations.

As Peirce's charismatic spell lifted following his death in 1880, historians of Harvard College began to dissect his modus operandi. As one alumnus grumbled, "He never seems to have grasped the essential difference between a sound mathematical proof and a true statement followed by a mass of mathematical verbiage which really proved nothing."[165] Peirce's confidence that he was authoritatively speaking in "the divine language" apparently led to a feeling of papal infallibility. Others highlighted a less obvious humility that was a byproduct of his mathematical idealism. As Peirce had written in one article, "When our researches are successful and when a generous and heaven-eyed inspiration has elevated us above humanity, and raised us triumphantly into the very presence, as it were, of the divine intellect, how instantly and entirely are human pride and vanity repressed, and, by a single glance at the glories of the infinite mind, are we humbled to the dust."[166] Obituaries from a wide range of secular and religious journals sainted him as a modest fighter for a faith that transcended the lines of human culture.[267]

These more generous accounts emphasized an important aspect of the mathematical work and religious belief of thinkers such as Benjamin Peirce and Thomas Hill: a fervent antisectarianism, a failure to perceive real distinctions between the various clans of worship. This robust ecumenism was a common element of idealist mathematical thought in the nineteenth century, a trait that would later ease away from its religious background to become the cosmopolitan pacifism still apparent in today's scientific circles. In America, land of religious freedom and forthright religious practice, the urge to transcend sectarianism through mathematics had only a modest and occasional impetus, however. Back in the British Isles, where religious differences entailed greater tension and upset, this motivation was much clearer.

CHAPTER THREE

George Boole and the Genesis of Symbolic Logic

George Boole had a revelation at the age of seventeen that changed both his life and the course of Western philosophy. He spoke in spiritual and mystical terms about the insight gained in this epiphany, which forever separated modern logic from the logic of Aristotle and the Scholastics. The mind, Boole discovered, has an innate sense of "Unity" that it constantly uses to synthesize its understanding of the world. Our natural propensity is not to see objects individually—as utterly distinct things—but rather in relationship to a greater, all-encompassing whole. In the mathematical terms most germane to the young Boole, there appeared to be something special about the number 1. Using that number to represent the "Universe of Thought" and variables to represent subsets within the whole, Boole eventually fashioned a new science of logic out of the symbols and processes of mathematics.[1] This mathematical logic, he strongly believed, could more effectively present and manipulate propositions, categories, and relationships than the overly wordy traditional system of logic.

Despite the transference of his emotional insight into the mechanics of reasoning, Boole never forgot that the notion of a mathematical logic came in the form of what he regarded as a divine manifestation, and he hoped to return the favor by applying his new logic to the advantage of religious belief. "The hope of his heart," his wife Mary would later recall, had been "to work in the cause of true religion." She continued, "Mathematics had never had more than a secondary interest for him; and even logic he cared for chiefly as a means of clearing the ground of doctrines imagined to be proved, by showing that the evidence on which they were supposed to rest had no tendency to prove them. But he had been endeavouring to give a more active and positive help than this to the cause of what he deemed pure religion."[2] The "father of pure mathematics," as Bertrand Russell would later refer to Boole, had not been purely interested in mathematics, nor was his mathematics free of the "impurities" of extradisciplinary concerns, in particular, religious ones.[3] The symbolic logic that is now an essential tool for secular philosophers and that forms the basis for dispassionate computers began in the mind of a warm-blooded, religiously concerned idealist.

In communications with his friends and former pupils, Boole spoke about the broader intent of his life's work more candidly than in his articles and books. In a telling 1840 letter, for instance, he wrote of the workings of the mind in relation to greater truths about the universe, and envisioned the study of thought in a manifestly religious way: "The ideas of human immortality, of modes of being infinitely diversified and bearing no relation to our existing senses in the present life of unlimited advancement and continued development, these . . . [are] among the glorious possibilities of the science of mind. And I hence am inclined to believe that the study of mental philosophy and the trains of reflection to which it naturally leads are favourable both to the growth of genuine poetry and the reception and appreciation of religious truth."[4] Just as William Paley and other natural theologians sought to bolster religion through the examination of nature, so Boole proclaimed that a sustained investigation of the mind would lead us to central religious tenets such as the immortality of the soul and the existence of a divine realm unavailable to our senses. Like the ingenious structure of the human eye or the perfect geometry of the nautilus shell, elements in our consciousness, such as the innate notion of "Unity," present compelling clues to the divine and beneficent construction of the universe. The human mind is no less of a miracle than the body, Boole conceived, its resplendence pointing to its creation by a higher power. In short, a turn inward would point us toward the external reality of God.

To understand the genesis of symbolic logic, therefore, it is necessary to understand the role of such nonlogical notions at work in George Boole's thought. The influence of certain philosophers and theologians is certainly an important part of the story. For instance, Boole was fluent in German and had read the works of Kant with more than a passing interest.[5] Yet just as important to the origin of Boole's new mathematical system are key elements of British society in the nineteenth century, especially religious conflict and the shifting winds of belief. Comprehending the meaning and significance of Boole's symbolic logic thus requires equal parts technical discussion, biography, and broader cultural history.[6]

An Environment of Conflict

The life of George Boole is one of the great success stories of the Victorian meritocracy. From an extremely modest family background—he was born the son of a cobbler in Lincoln in 1815—he rose to prominence in a number of fields. His father, John Boole, was self-educated and especially fascinated by mathematics and astronomy, and he inculcated these interests in George. When he was a young boy, Boole's father would reward him with a treat for studying Euclid.[7] Much like his contemporary John Stuart Mill, George Boole was a child prodigy, mastering languages both ancient and modern. He also devoured every mathematical treatise he could get his hands on. By his late teens, having never attended university, Boole started and ran his own school while working at night on numerous mathematical advances based on the work of Laplace and the Cambridge Analytical Society. Members of that mathematical group eventually wrote effusive letters of recommendation that secured him a position at the nascent nondenominational Queen's College in Cork, Ireland, in 1849. From there, Boole's international fame as a mathematician and logician grew markedly, especially after the publication of *The Laws of Thought* (1854), and he cemented his newfound status through a marriage to Mary Everest, the niece of George Everest (the surveyor-general of India after whom the mountain is named). George Boole no doubt would have gained even more prominence had he not walked miles on a chilly, raining day to attend a lecture on mathematics. He caught pneumonia and died in 1864.

Throughout his life, Boole found himself constantly on the run from political and religious factionalism. This pattern began with his very first experience as a teacher, at a Methodist school in Doncaster in the early 1830s. Local adherents became displeased when they discovered that Boole was not Methodist himself, and they began to pray for his conversion. Boole, not wanting to become embroiled in a sectarian dispute, quickly left for another post.[8] For years a peripatetic teacher, he saw such partisanship everywhere he looked and wondered why others were not as appalled by it as he was. In 1848, Boole relished how the shock of the Continental revolutions upset the generally serene sphere of academia, and he believed that factionalism would finally have to be addressed in the ivory tower. Writing to his friend Charles Kirk in the spring of 1848, Boole seemed delighted that even the "pale young men" of Cambridge, "intent upon Greek accounts and double integrals," would now have to "lift up their heads and speculate upon the world whose existence they had almost forgotten."[9] Boole himself read the revolutions of 1848 as a complete and utter failure of Christian brotherhood, and yearned for a more peaceful day: "It is most earnestly to be desired that Christians could be brought to discharge their obligations to society without foreign excitement; but I suppose that we are not to look for this before the millennium."[10] Boole saw the ubiquity of conflict in Europe as a damning summary of the age: mankind hopelessly unable to coexist peacefully, without divisions. In verse, Boole expressed his woe over the revolutions of 1848:

> Oh, too long sever'd in the thought and speech,
> Of mortal men, too oft as rivals set
> Who kindred are, and in firm union met
> One consummation in the Heavens shall reach.[11]

He pined for "another day" when "the reign of Wrong is past for ever/Because a higher Law to us hath spoken/And war is not the height of man's endeavours."[12] An apocalyptic cleansing of humanity and a rebirth of brotherly love seemed distant, however, in the shadow of 1848.

Boole nevertheless viewed his appointment to Queen's College, Cork (QCC) as a good opportunity to advance unity, and he had high hopes. He went to Ireland as a young, proud founding faculty member of the college. At a public testimonial held in his hometown of Lincoln before he left for Ireland, Boole rose to address the audience with a great sense of purpose and idealistic vision. A local newspaper recorded that Boole saw one of the most pressing goals of the Queen's University system to be "the bringing together and associating of the Catholic and the Protestant youth," and he was confident that the new colleges "would contribute to the harmony of Ireland, and that they would in their internal management prove models of that peace and harmony which they recommended."[13] The new college seemed to provide a window of opportunity, bringing together disparate people in the common interest of education and progress. For Boole, therefore, QCC was a testing ground for the ability of various factions to get along in learned coexistence, and he placed an enormous personal and religious stake in the outcome. Could it pave the way for a deep and true reconciliation between two lands and two religious sects that had a long and sad history of discord?

Upon arrival in Ireland, the possibility of harmony suddenly seemed less likely as Boole witnessed firsthand the Great Famine as he traveled through the Irish countryside on his way to Cork. The disquieting and omnipresent poverty in Ireland clearly occupied Boole's thought in the inaugural school year of 1849–1850, leaving a lasting impression on him. Living on the river in Cork, he observed in horror how the spring flooding of 1850 took a grave toll on the cellar-dwelling poor.[14] Boole was even moved to write poetry about the dismal situation, an act he always reserved for matters of utmost significance. An empathetic cry begins a sonnet he wrote in November 1849:

> O Ireland! Ireland! though thy tale be told
> On history's darkest page, and to such strains
> Of Sorrow's harp attuned, that there remains
> No chord of grief untouched, though Want be bold
> And clamours in thy streets, and where the gold
> Of plenteous harvests waved, lie plashy plains,
> O'er which the bulrush towers, the ragweed reigns.[15]

The difficult conditions of the famine heightened Boole's sense of urgency that something had to be done to pull the inhabitants of the British Isles together.

Because the famine had magnified Irish nationalism, however, Boole quickly found himself largely alienated from the surrounding populace. His feeling of otherness was validated when he attended a function in March 1850 at the house of a government official. In a long letter the mathematician wrote to his mother afterward, he expressed astonishment at how his Irish counterparts voiced disconcerting views about the aims and character of the British people and their government. "It is almost inconceivable what notions some of the Irish middle and upper classes entertain of England," Boole complained, "You would imagine from their talk that they think . . . that England delights to see Ireland miserable and desires nothing so much as to keep her so." Boole tried to discount some of the grumbling as an "exaggerated way of talking" in Ireland, but he was clearly put on the defensive by the other guests.[16]

Boole nevertheless remained sanguine about the chances for his Queen's University branch during the first year of classes. He made friends quickly, an achievement reassuring for someone with a somewhat shy demeanor. Another factor that helped ease Boole's entry into the predominantly Catholic country was Cork's substantial Protestant population. In one of his first letters back to England, Boole even described himself as "very comfortable here," and thought that "at present everything seems to promise harmony."[17] Furthermore, QCC appeared to be situated in a relatively moderate part of the country, and the citizens of Cork seemed to possess a great reserve of tolerance. "I have nowhere met with people so kind, a few individuals excepted, as in this city," Boole relayed to his sister in England in May 1850.[18] Boole's first impressions of Cork were quite positive, though he worried that "if that charity which was the essence of true religion was lost amid the turmoil of theological disputation," the college would sink quickly in a sea of controversy.[19]

This conjecture turned out to be quite prescient, as the Queen's University experiment quickly deteriorated in an environment of growing sectarian sensitivity. Although now an integral part of Ireland's higher education system, at their inception the Irish viewed the three Queen's Colleges with great suspicion. Minority factions such as the Methodists supported the Queen's Colleges almost immediately, but many Catholics judged the schools as an attempt at what today would be called "cultural imperialism." Although British Prime Minister Robert Peel claimed that higher education would help to raise Ireland from its depression, budding Irish nationalists questioned Peel's true motives. In addition, Pope Pius IX spoke of founding a competing Catholic university.[20] Nevertheless, numerous moderate Irish Catholics accepted the colleges (perhaps grudgingly), and Catholics comprised fully half of the entering class of the Queen's College in Cork in 1850.

A significant extremist Catholic faction, however, vowed to eradicate this new British invasion. These vociferous opponents obviously made themselves heard more than the silent moderates, and they frightened the imported English professors and administrators of QCC, including Boole, by early 1850.[21] In the summer of that year at the Synod of Thurles, Catholic antagonists of QCC found a strong voice in Paul Cullen, the newly elected Primate of Ireland. In a rabble-rousing speech, Cullen ominously warned of "the propagation of error through a godless system of education."[22] The Catholic hierarchy in Cork heard his words clearly. At the beginning of the second school year, in the fall of 1850, Boole related a common tale: "I saw one [Catholic] student yesterday who told me that the college had been denounced from the altar in his parish and that he was afraid to return accordingly."[23] Indeed, in their Sunday sermons Catholic priests condemned QCC and castigated the college's students with increasing frequency.

The anti-QCC faction found grist for their mill when Louis Raymond de Vericour, a professor of modern languages and a housemate and friend of Boole's, unwisely published a highly controversial work entitled *An Historical Analysis of Christian Civilisation* in 1850. The book appeared to make several assertions that undermined Roman Catholicism. For instance, detractors could easily read Vericour's claim that "nothing satisfactory is really known" about the Bishop of Rome's "assuming of pontifical authority" in the early Christian era as a strike against such authority.[24] In general, the book's strongly progressive vision of history colored major events such as the Protestant Reformation in a far more positive light than the sections on the Roman Catholic Church. In many passages the Church came across as a stodgy, overly conservative antagonist of truth.

For Catholics opposed to the Queen's Colleges, *An Historical Analysis of Christian Civilisation* seemed a gross affront to their sovereign faith and an indication of the underlying disposition of the new university. As the affair spilled out into the newspapers and became a much-discussed topic among the residents of Cork, QCC's administration suspended Vericour under the institution's rule that all professors maintain religious impartiality.[25] However, he appealed his suspension on the grounds that the nondenominational charter of the Queen's University only mandated neutrality on religious issues in the classroom, or when acting as a representative of the university. What professors said in their publications or did on their own time could not count against them, Vericour argued. It was a shrewd legalistic appeal, and he won his job back. Eventually the matter subsided, but the Vericour affair was not what the supposedly evenhanded school system needed to bolster its fragile reputation.

Meanwhile, simultaneous events in England exacerbated the disputes surrounding Queen's College, Cork. As Boole was heading off to Ireland at the end of the 1840s, thousands of Irishmen were passing him in the opposite direction. Combined with the revived Anglo-Catholicism that arose out of the Tractarian movement of the prior decade, the Catholic population in England grew twentyfold between 1800 and 1850.[26] By that year Pius IX thought it appropriate to reestablish an ecclesiastical hierarchy in England. In the late summer of 1850 he did just that, appointing a new set of bishops and a cardinal. Strong anti-Papist and anti-Catholic feelings quickly surfaced. In the minds of most British Protestants, Anglo-Catholicism—previously informal and isolated—instantly had become manifestly bound to Rome. Violence against priests and "No Popery" riots ensued, peaking on Guy Fawkes Day, November 5, 1850. A letter from Lord Russell released to the press two days later fanned the flames. Russell inveighed against the encroachment of Rome and also used the occasion to impugn the Tractarians (who had always been set against him) as insidious double agents for a foreign power. Russell, who had continued to give significant support to Peel's Queen's University initiative, now appeared to most Irish Catholics as a sworn enemy of their faith. In turn Russell used the Synod of Thurles's unbridled attack on the new educational system as further evidence of Rome's aggressive stance. The battle lines had been drawn, and anti-Catholic action by the British Parliament—reversing two decades of liberalization—seemed inevitable.[27]

In the eye of this sectarian storm were the Queen's Colleges, and Cork had the unhappy fate of having a poor captain at its helm. Always generous in his assessment of human beings, Boole had found the first president of QCC, the Irish Catholic chemist Sir Robert Kane, agreeable enough when they first met in the fall of 1849. The mathematician told his mother that Kane "is I think from all that I see of him a very kindhearted man," and he expected to have an amiable relationship with the president.[28] However, even the slightly naïve Boole understood that these were first impressions, and as soon as the president began to speak out and make decisions his image changed considerably.[29] Kane's most significant blunder was his presidential address to the students, faculty, and staff at the start of the school year in the tense fall of 1850. Entrusted to maintain the nondenominational character of the new college, Kane promptly diverged from universal themes into a discussion of hot-button sectarian concerns. Many faculty members, especially Protestants, found this more than a bit disconcerting and implored the president to honor the founding principles of the college. Boole was among those who were troubled by Kane's address: "It was generally thought by the professors to be much too exclusively occupied with Catholic questions to the almost total exclusion of another faith either in the college or out of it. I believe that he will get a hint that the principle of religious equality must be respected in appearance as well as in fact."[30] Kane unfortunately did not get the hint, even after repeated faculty admonishments, including one uneasy meeting with Boole.[31]

The uproar over the events of 1850 and the QCC president's reaction to them obviously disturbed Boole and other professors, who worried that the sharpening of doctrinal divisions would traumatically splinter the college. Boole wrote to Augustus De Morgan in October 1850 greatly lamenting "the storm of religious bigotry which is at this moment raging around us here."[32] For Boole, the intensification of sectarianism represented his worst nightmare of religious and political discord. In addition, he worried that the conflict might sink QCC. "The ill feeling which is springing up between Protestant and Catholic in England will be likely to affect our success here very much," Boole nervously wrote his sister in November 1850.[33] In the winter of 1850–1851, Boole recorded that some anti-Protestant extremists were "more violent than ever," but added in a letter home, "I see nothing of them and only judge from an occasional sight of their newspapers . . . nor has there been much hostility between the establishment and sects of Protestant dissenters here so that matters are on the whole more quiet and peaceful than one could have expected."[34] A tenuous détente somehow materialized.

Yet the antipathy and invectives had begun to take their toll. The QCC student body shrunk considerably following Catholic sermons emphasizing the impropriety of attending a college created by Protestants.[35] Boole started to worry about his class size, and with good reason—professors' salaries were based on how many students they taught. Having to support his mother, his sister Maryann, and often his brother Charles, stretched George's already meager earnings. Upon hearing in January 1851 that his income for the 1850–1851 school year would be a mere £290, he bemoaned to Maryann, "This is too little. I am afraid the prospects with us are not brightening."[36] Boole realized that since he taught relatively esoteric topics his Queen's College salary was not likely to rise above the modest sum of £350.[37] These financial concerns, directly related to the more weighty subjects of religion and politics, weighed heavily on the young mathematician's mind.

When President Kane began to meddle in the teaching methods and syllabi of the professors, matters only got worse. Though John Ryall, the vice president of the college, allowed Boole free reign in the department of mathematics, Kane publicly questioned Boole's program of study. He advocated the teaching of more practical mathematical subjects, whereas Boole believed strongly in the superiority of abstract topics and methods. A lengthy and often public battle between the two ensued, with each writing accusatory letters to local newspapers. Other QCC professors experienced similar challenges to their professional and pedagogical authority. Boole envied a professor of geology who had escaped to a chair in Aberdeen in 1853, where the mathematician presumed his former colleague would "enjoy peace and freedom."[38] What galled Boole most of all was Kane's unpredictability and unwillingness to listen to the calm voice of reason. "When the most reasonable and temperate efforts to bring about a better state of things expose a man to the charge of faction and subject them to the frown of power, I do not see what but ruin can be expected," Boole complained to a friend.[39] Kane seemed irreversibly illogical, unable to maintain the necessary impartiality required for his position. To Boole, this partisanship was Kane's worst sin, and to be accused of such in return—when he believed himself to be a crusader against all forms of factionalism—represented the height of injustice.

More important than Kane's character and miscues, increasingly frequent disputes in the 1850s between Catholics and Protestants destroyed any hope for accommodation at QCC. The smallest events could touch off a firestorm of protest. In 1855, for instance, Church of Ireland laymen in Cork became infuriated and demanded an explanation when their bishop invited the Catholic bishop to a dinner that the Lord Lieutenant planned to attend also.[40] Religious lines even extended into the secular realm of the public house. In 1856, the Catholic Murphy brothers began producing their Irish Stout to compete against the dominant—and Protestant-owned—Beamish.

The final blow to QCC occurred when the Roman Catholic Church officially condemned the Queen's University. Priests' individual denunciations escalated into a concerted effort by the ecclesiastical administration to renounce the "colonizing" educational system. In the spring of 1857, Archbishop MacHale convinced the Pope to condemn the Queen's Colleges and fund the construction of Catholic schools. Thereafter, pressure increased tremendously on QCC's Catholic students to abandon the censured college for the new Catholic University. In 1857, Boole summarized the state of affairs to his sister: "What can [the Catholic students of QCC] do when their church commands[?] They must obey or quit her communion . . . It will bring matters to a crisis."[41] He feared the worst, seeing the formal papal condemnation as just the beginning of the Roman Catholic Church's undermining of nondenominational education. More "thunder" was certainly on the horizon and the end of ecumenical dialogue in a protected sphere—the college campus—was near.[42]

Surprisingly, during this escalating factionalism Boole felt little compulsion to return to an academic position in England. His homeland seemed just as mired in sectarian discord as Ireland. As early as November 1850, Boole turned down a professorship in Manchester, apparently because of religious concerns.[43] The thought of sailing back to England during the anti-Catholic riots made his uncomfortable situation in Cork slightly more palatable. Later in the 1850s, a position in mathematics opened up at Oxford University. It was a plum job, a position Boole had coveted for decades. He envisioned the Oxford professorship as a chance "to develop the teaching of mathematics as a healthy, moral discipline."[44] Boole was hesitant and even somewhat skittish about applying, however. As his wife later recalled, his presence in Oxford meant that "he would be expected to take one side or the other in [the] theological controversy" between High and Low Churchmen. "The life of a man who would be a partisan of neither side might be made very uncomfortable," Mary Everest Boole explained.[45] Ultimately, George Boole equivocated. He halfheartedly sent his name in as a candidate, but the application lacked any pretense of sincere interest and he failed to submit supporting testimonials. The job went to someone else.

Boole never returned to England; he also never fully integrated into Irish society. Mourning his move from Lincolnshire to Cork, Boole cried in the voice of Britain, "What has that Hibernia done/Thus to steal my favourite son?"[46] In 1860, a decade after his arrival on Irish soil, the mathematician summed up his life in Ireland to his friend A. T. Taylor. Boole lamented that he lived "in a country in which I can never feel at home," and complained that "the Roman Catholic Priesthood seem to have been doing all they can to preach disloyalty. Between them and a bigoted Calvinistic Protestant population this is a country which does not on the whole present the most favourable picture of Christianity."[47] For his entire adulthood, George Boole felt like a man without a home on a map divided by lines of religious and national identity.

Boole's Spiritual Development

In this context of ubiquitous and seemingly intractable discord on the local, national, and international levels, George Boole looked to his faith for salvation. Like so many of his fellow Victorian intellectuals, however, Boole could not find solace in a simple, by-the-Book Christianity. His spiritual biography consists instead of a journey from a fairly straightforward association with the Church of England to a nonconformist faith based on religious idealism. Along the way Boole dabbled in increasingly broader and more liberal forms of Anglicanism, seriously considered Unitarianism, and perhaps even flirted with Judaism. His *Bildungsroman* was similar to the spiritual trek taken by John Henry Newman's brother Francis Newman (1805–1897), a mathematician and classicist who also happened to be Boole's good friend. Boole agreed with Newman's assessment in 1850 that "Practical Devoutness and Free Thought stand apart in unnatural schism. But surely the age is ripe for something better;—for a religion which shall combine the tenderness, humility, and disinterestedness, that are the glory of the purest Christianity, with that activity of intellect, untiring pursuit of truth, and strict adherence to impartial principle, which the schools of modern science embody." Like Newman, Boole wished to "lay aside disputes of words, eternal vacillations, mutual illwill and dread of new light," and move on to a new era of spirituality that would heal the age's terrible divisions.[48]

Unlike Francis Newman, who (like his brother) published a spiritual autobiography, Boole's aversion to criticism made him ever more reserved about expressing his creeping nonconformity. After one such disclosure in the mid-1840s, for instance, he had to speak at length with a Lincolnshire rector to defuse growing concern about his religious belief. "I hope this will redeem me from the stigma of unfaithfulness," Boole wrote to another clergyman.[49] Ultimately, however, the mathematician could not conscientiously stay within the confines of the Anglican communion, and the hallmark of his faith became a robust idealism divorced from institutions and dogma. This religious idealism became a primary animator of his work in mathematics and logic.

George Boole undoubtedly began his religious life with few signs of this later spirituality. His parents inculcated a strong Low Church Anglicanism, the young boy read the Scriptures frequently, and from his early teens he desired to become a minister in the Church of England. As he prepared to take the holy orders in his late teens, however, his parents' financial state deteriorated and he had to abandon his religious career to support his family by teaching.[50] Nevertheless, Boole was able to satisfy some of his ministerial yearnings by including religious instruction in his program of study. He used fairly standard Protestant textbooks, including one consisting of doctrinal questions followed by answers based on biblical passages, and another comprising New Testament biographies.[51]

But as Boole grew older his attachment to the New Testament and the figure of Jesus Christ decreased significantly. For example, when the controversy over David Friedrich Strauss's less-than-transcendental historical portrayal of Jesus arrived in Great Britain, Boole expressed little angst. As far as he was concerned (he later told his wife and children) the historical Jesus was completely insignificant compared with the *idea* of him. Questions raised by Strauss and other "higher critics" of the Bible, so alarming for most Christians, seemed ultimately irrelevant to Boole. How could one care so much about Jesus' biography when what really counted was the moral ideal Jesus represented?[52]

Family and close associates recognized that after the age of twenty Boole gravitated toward Unitarianism. Even as early as Boole's tenure at Waddington Academy in the mid-1830s, one student recorded that the teacher appeared far truer to Unitarianism than Anglicanism.[53] Students at his school in the late 1830s recalled that Boole had drifted from the Church of England, reading from the Greek Bible rather than the authorized Church of England version.[54] Regarding the Trinity, Boole told his wife that if the doctrine "were in any sense true, it must be in this sense: Not that there are three Persons in the Godhead, but that man is so constituted as to be capable of comprehending three, and only three, modes of manifestation of God."[55] His sole interest in the Trinity had less to do with the nature of the Father, the Son, and the Holy Ghost, than with the apparent specialness of the number three itself, especially given that human beings perceive the world in that number of dimensions.[56] Although Boole sometimes went to Unitarian services, and even thought about committing himself to the sect, ultimately he could not bring himself to endorse the Unitarian dogma either.[57]

Like many of his contemporaries, Boole eventually had a full-blown crisis of faith. In a rare straightforward expression of his state of belief, Boole admitted to a confidant in 1840 that he was having trouble reconciling himself to Christianity. In the first half of the letter he wavered, feeling hesitant "to avow myself in belief a Christian," yet placing his "hopes of future happiness on the great propitiatory sacrifice and . . . merits of the savior." Boole nevertheless concluded, "I doubt whether I am a Christian at all except in mere speculation" and he pleaded with his friend not to press the matter any further.[58] Indeed, Boole was already teaching his students religious notions far removed from Christianity. In one science lecture at his school in Waddington, Boole startled his student Charles Clarke with a discourse on electricity in which he described the phenomenon as "the great motive and life-giving power . . . the source of heat, of life, of motion . . . the root of all, the beginning of all ... the one natural agent by which the creator performed his great work in the universe."[59] Boole also delivered a lecture entitled "Are the Planets Inhabited" that answered in the affirmative. Since God works by "fixed and immutable principles," and since He seems to have constructed planets for the purpose of life, they must be inhabited, the mathematician concluded.[60] Needless to say, parents were not amused that Boole was subjecting their children to such speculations.

Another element of Boole's early education obviously was coming to the fore: his father's emphasis on the laws of the universe and the sciences that allow humanity to comprehend them. In addition to promulgating Anglicanism, John Boole had highlighted the majesty of astronomy to the young boy. Self-taught in mathematics and the physical sciences, the elder Boole built his own telescope and used it to teach his son about the patterns of the heavens. A sign in John Boole's shop window read, "Anyone who wishes to observe the works of God in a spirit of reverence is invited to come in and look through my telescope."[61] George Boole, in turn, passed on this awe of nature to his students. Charles Clarke recounted a typical story: "When I was a little boy, he took me, one Sunday afternoon, along the Greetwell Fields ... The Ministege Bells were ringing in for afternoon prayers and I told him that we should be late for Church. He smiled and told me that we were in Church, a church built by God himself and not by men, and asked me whether God could not be worshipped as well among the trees and in the fields as in houses made of stone?"[62] Boole told Clarke and other students that the beauty and wonder of nature, like the vaulted ceilings and spires of a Gothic church, led one upward to the contemplation of God. In a sonnet on springtime, Boole spiritually joined the magnificence of nature with the revelations of Galileo:

> The highest truth which in our hearts we hold
> Makes of itself in outward things a sign,
> As in the vision'd stream through flowers which rolled,
> Heaven's courts appeared to the great Florentine.[63]

Many of Boole's surviving poems are similar odes to "God's handiwork," emphasizing the ascendance from nature to the divine realm.[64] This procession upward is critical, for Boole—like Benjamin Peirce and Thomas Hill—did not wish to focus too much on material nature itself. The supreme end of worship, he believed, should be nature's governing Mind. In an 1847 address on the solar system and stars, Boole concluded with a telling exclamation: "What indications of order and presiding intelligence throughout the whole!"[65] God, he asserted, was to be found in these higher, perfect notions of order and intelligence, not in the actual structure of the cosmos.

Since Boole thought that the realm of transcendental ideas, not the physical world, ultimately provided the true path to God, the mathematician unsurprisingly began to scrutinize the human mind in the 1840s. As he wrote to a friend before commencing his examination of logic, "I conceive it impossible that an individual should look with much fixedness of attention on the phenomena of this inward life and being without the feeling that each observed fact, each ascertained truth, is but one link out of an infinite chain of possible truths of which each may afford matter for more sublime contemplation than can be derived from any of the forms of material grandeur and beauty."[66] Boole believed strongly that this mental investigation would lead to a greater comprehension of the character of God than any notions arising from natural philosophy.

Boole's encounter with Immanuel Kant's philosophy in the 1840s solidified his feeling that God existed on a plane beyond space and time. Following Kant (whom he was able to read in the original German) Boole conceived the universe as split into two realms: the phenomenal realm of time, space, matter, and the senses, and the noumenal realm of God, the human soul, and the transcendental aspect of the mind. Boole wrote to Augustus De Morgan in October 1850 agreeing with Kant that "there is a real phenomenon in the mind . . . which distinguishes it from the system of external Nature."[67] This precious mental gift gave human beings the opportunity to rise above the base material world and participate in the noumenal realm. By the late 1840s, Boole's poetry also displayed Kant's influence. In one sonnet he described the twofold nature of humanity: "We / In whose mysterious spirits thus are blent / Finite of Sense, and Infinite of Thought." Boole then wrote in reverence about the higher faculty of the mind, which allows us to peek "behind the veil phenomenal" and discover the heavenly realm.[68]

More generally, Boole repeatedly praised the invisible and disparaged the visible. Meditating on a passage from the Bible, for instance, he emphasized that "the glories of this visible world," "the glories of these visible heavens," and our existence in the form of matter, "shall pass like dreams away." Mankind should instead strive for the everlasting and invisible sphere of "Truth," which we can find by looking inside of ourselves and using the transcendental mental faculty given to us by God.[69] In Boole's verse, the most religious individual was thus the

> Seeker after Truth's deep fountain
> Delver in the soul's deep mine
> Toiler up the rugged mountain
> To the upper light divine.[70]

As he once told his wife, God is by nature invisible and any attempt to base religion on a "visible manifestation" of Him is preposterous.[71]

Yet Boole, much like the American idealists, also blurred Kant's distinction between the visible and invisible. Whereas the German philosopher strictly separated the religious faculty of the mind, which rose to the ethereal realm of God, from the scientific faculty, which only dealt with the world of matter, Boole claimed that science and mathematics could be transcendental as well. As he declared to an audience in 1847, "The great results of science . . . have an existence quite independent of our faculties and of our recognition."[72] Although Kant categorized the axioms of geometry as innate ideas, that is, as part of a framework the mind imposes on the external world, Boole thought these concepts originated in the noumenal realm. Mathematical laws were not products of the human mind but rather transcriptions of the mind of God. A favorite verse of Boole's, "For ever O Lord, Thy word is settled in Heaven," was for him a description of mathematical equations.[73] Though the divine ideas of mathematics reside in heaven, advanced thinkers could, with great effort, command them. This mastery of divine principles was what made the discovery of Neptune so special for Boole—Adams's and Le Verrier's use of the laws of geometry meant that their work was a religious triumph as much as a scientific one.

Indeed, the discrete adjectives "religious" and "scientific" made no sense to Boole—truth seeking was simultaneously both. The pursuit of scientific truth was also a religious act. According to Mary Everest Boole, her husband "interpreted such expressions in the Psalms as 'The word of God,' 'The testimonies of the Lord,' etc., as meaning . . . 'all the laws of the universe.'"[74] Displaying his strong Platonist bent, spirituality for Boole entailed being "in contact with truths that are eternal," and this notion allowed him to write about seemingly dry mathematical topics with a fervor akin to religious rapture.[75]

Boole's nonsectarian—indeed, deeply antisectarian—faith thus claimed that divine truth was singular and transcendental. The varied, conflicting doctrines of Low and High Churchmen, Quakers, Unitarians, and Catholics all dissolved in the unity of God's mind, into which Boole desperately wished to escape. Although this idealism led the mathematician away from traditional Christian dogma and liturgy, he still saw himself as loyal to what he saw as the core of that religion. Boole agreed with his contemporary, the logician and theologian Richard Whately, that the distinctive aspect of Christianity was its assertion that God "regards all as alike, acknowledges no caste, espouses no system of reserve."[76] If Boole remained a Christian, it was in this liberal sense, and he expressed it both in his social interactions and, perhaps most strikingly, in the development of his mathematical logic in the late 1840s and 1850s.

Interfaith Relations

George Boole had a particularly insidious toothache the day he was introduced to the Roman Catholic and Protestant bishops of Cork. His face was not its normal long and sharply-angled self, but rather swollen and rounded. It just so happened that the Catholic bishop was renowned for his portliness but the Protestant bishop was relatively thin. The Catholic bishop looked at Boole and saw a Protestant; the Protestant bishop looked at the puffy Boole and saw a Catholic. Boole himself understood the moment as a metaphor—like a chameleon he would blend into different religious groups.[77] Though nominally a Protestant for most of his life, Boole's religious idealism propelled him into regular association with members of other Christian sects and other faiths. He was a strong supporter of religious toleration and adapted himself easily to many different communions and liturgies. As his wife recalled, "George seemed never tired of reading what are called 'religious books.' Hymns, the works of Eastern and Christian mystics, sermons, Jewish books of devotion, the works of old Fathers and Puritan divines, interminable discussions on the Atonement, the Trinity, the nature of Revelation, the lives of saints, Catholic and Protestant—he had always some of these on hand."[78] The mathematician was a religious omnivore, and the ease with which he engaged in intersectarian and interfaith communication made him a natural and eager mediator.

Boole maintained important relationships with believers of all stripes. On a given Sunday, he could be found in any one of several houses of worship. Drawn like a child to the piper, Boole would attend any religious service that had a modicum of music, whether in a Protestant, Catholic, or independent church.[79] He even visited the Society of Friends on occasion, and sustained a lively intercourse with a Quaker in the Cork countryside.[80] Because of his peripatetic style of worship, Boole declined a generous offer from the Archdeacon of Cork—the use of his private pew whenever Boole wished to attend services at his church.[81] Adhering to the idealist creed, it did not matter where or with whom Boole worshipped; in his view, no sect had a monopoly on God. As a young man, Boole captured his beatific (and distinctly British) vision of interfaith relations: "The religious world is growing very social and very fond of tea which are two good signs and which auger well for a future age. Very probably, if tea had been introduced into Europe at an earlier period, it might have done something for the irritations, and diminish the acrimony, of religious differences."[82] Boole steadfastly practiced this tea-time affability, a temper he hoped would return humanity to a single communion.

Boole admired those who were able to rise above divisions and join people in faith. A prime example for him was Thomas Arnold (1795–1842), the headmaster of Rugby. Boole disagreed with Arnold's theological positions, yet commended the way in which he placed brotherly love squarely before dogma. Recalling Arnold in 1845, Boole wrote, "Though a churchman Arnold was of a truly catholic spirit—though warmly attracted to his principles and fearless and unshrinking in the avowal of them . . . yet was he a lover of peace, a man of truly gentle and affectionate mind . . . no man so slandered and castigated during his life has so enlisted after his death the suffrages of the sincere and good of all parties."[83] After reading Arthur Stanley's biography of Arnold (1844), Boole dreamt of a world in which a new Church, "throwing aside the vain dream of uniformity, should allow different forms of ritual and tolerate a diversity of opinion."[84] Most of Boole's closest friends were Arnolds—deeply faithful, yet open-minded, humble, and extremely tolerant of those with viewpoints different than their own.

Unthinking subscription to religious tenets was one of Boole's greatest *bêtes noires*. "Once a man thinks himself bound to a settled creed, it seems as if truth, faith, and charity become impossible to him, except in so far as he evades his creed," he told his wife.[85] In this spirit, Boole avoided attending church on Sacrament Sunday to remain neutral with respect to the dogma of the Eucharist.[86] Boole's definition of the sin of "idolatry"—"attachment to any particular doctrine"—underlined his antidogmatic stance.[87] A major problem facing mankind, he thought, was how most people, including those who were well educated, were overly inclined to accept religious propositions without full consideration.[88] One of Boole's poems from the mid-1840s asked readers to question whether they had really followed the truth and way of God, or had instead upheld false premises:

> Hast thou with higher emulation lighted
> Shrunk not thy fearless footsteps to advance
> Amid the deeper gloom of souls benighted,
> The moonless, starless night of Ignorance,
> With Truth's undying torch the gloom dispelling
> From many a hopeless, many a darksome dwelling.
>
> Hast thou thine intellectual being nourished
> With the pure draught of truth's perennial stream?
> Or clasped the darling prejudice, and cherished
> The passionate illusions of a dream?[89]

Religious doctrines can appear unassailable within one's world view, Boole emphasized, and so finding true faith, rather than blindly adhering to dogma, was a difficult task.

Boole reserved his rare angry outbursts for believers who placed worldly authority above the heavenly authority of God. For example, the mathematician differentiated between two types of Roman Catholics: those who he believed were married to the ecclesiastical hierarchy and power, and those who seemed to care more for the founding principles and values of the Church. "What is most desired in reference to the Church of Rome, is that an entire separation should take place, between the liberal and the ultramontane portions of her members," Boole wrote to his friend Joseph Hill in 1850, "The former are friendly to civil government, to mixed education, to freedom of conscience, and I believe not hostile to the Bible. Were they separate from the others they would approximate more and more the Protestant sect in discipline, in character, and doctrine. The members of the ultramontane party, on the other hand, are as I conscientiously believe, the enemies of all that is good."[90] Boole unsurprisingly avoided the "ultramontanes," but he developed lifelong friendships with the "liberals." Of one "liberal" Roman Catholic, James Simkiss, Boole affirmingly remarked that "he has as much of the Catholic and as little of the Roman about him as any man I ever met within whom the two terms were united."[91] One of Boole's favorite works was Richard Whately's *Errors of Romanism* (1830), which detailed point-by-point the supposed injustices of the Roman Catholic Church.[92] In other words, the mathematician admired the religiosity of faithful Roman Catholics while denigrating what he considered to be the improper and unholy jurisdiction of their Church.

Boole did not restrict his venom to the Roman Catholic priesthood alone; a virulent anticlericalism permeated his thought and writing. He found most clergymen ignorant, immoral, and unacceptably conservative, untrue to crucial precepts of their faith. The notion of Christian brotherhood, Boole told his wife, seemed alien to the Anglican clergymen who made off-color remarks about the Irish in casual conversation with their parishioners.[93] Although almost always soft-spoken and reserved, he would become enraged when discussing how so many clergymen opposed the advances of science.[94] In one instance, when Mary Everest Boole asked her husband to clarify what a pompous clergyman had meant in a sermon, George Boole replied snidely, "Means! . . . oh! [he] means—nothing; you mustn't suppose people *mean* anything when they speak in that way."[95] Perhaps most tellingly, one of George's final requests was that Mary not let the clergy take part in any way in the raising of their children.[96]

Boole's antidogmatism and anticlericalism, combined with his characteristic reticence, entailed that he often lived vicariously through those who challenged religious authorities. One of his best friends and favorite preachers was Edmund Roberts Larken, the Rector of Burton-by-Lincoln, who gained notoriety for being the first Anglican clergyman to wear a beard in the pulpit. This flaunting of ecclesiastical standards was relatively modest, however, compared with Boole's favorite rebel, John William Colenso (1814–1883), the Bishop of Natal. Colenso publicly criticized what he saw as the harshness of many Anglican ministers and officials. Situated among the Zulus of southern Africa, he found it absurd that Church of England missionaries were emphasizing doctrines such as eternal punishment rather than more fundamental (and gentler) precepts of Christianity. Furthermore, in the early 1860s Colenso published an extended literary analysis of the Pentateuch—not in the interest of pulling apart the textual roots of Christianity, but of making the religion less dependent on a rigid interpretation of the Bible.[97] Boole relished from afar Colenso's offensive against Church of England authorities and dogma, and as financial and ecclesiastical support for the Bishop of Natal quickly waned, the mathematician envisioned Colenso as a martyr for the true faith.[98]

Although Boole admired Christians who assailed sectarian rigidity, he perhaps most honestly revealed his impulse to transcend religious lines in his attraction to Judaism. Many contemporary Europeans despised the Jews for belonging to no nation—and thus, as several nineteenth-century ideologies held, for being "parasites" on other peoples and nations—but Boole saw this lack of a homeland as a sign of Judaism's ecumenical nature. While secular Positivists and many Christian theologians asserted that the Jewish faith had been superseded in the course of history, Boole saw Judaism's age as a positive attribute, not a negative one—if Judaism originated long before ubiquitous national and sectarian divisions, then perhaps it approached the purity of faith and true allegiance to God's morality Boole sought. In other words, in his desire for a unified faith Boole reversed much of the anti-Semitic logic of his day. His romantic vision of the Jews saw them not as contemptible relics of the past but as honorable representatives of an uncontaminated religiosity that existed millennia before factionalism.

Boole first expressed yearnings to understand the "presectarian" religion of the Jews in his quest to learn Hebrew. In his late teens and early twenties, he tried to acquire the ancient language with help from a Jewish mentor and friend in Lincoln, who also evidently assisted Boole in understanding the nature of his boyhood revelation regarding "Unity."[99] Despite tremendous motivation, however, Boole failed to learn Hebrew because of his poor eyesight and difficulties with the ambiguous characters and grammar.[100] Nevertheless, as a young man he continued to enjoy reading and studying the Old Testament more than the New.[101] As an adult, Boole moved on to Jewish history, medieval poetry, and contemporary newspapers.[102] He concluded that he was a kindred spirit of the Jews, because the Jewish people had suffered (on a larger scale) from the same religious persecution as himself. The mathematician felt a great deal of sympathy reading the story of a Jew who leapt into a fire rather than convert to Christianity.[103] A posthumously discovered box of manuscripts underscored Boole's strong affinity toward Judaism. "There was not a line in any of [the manuscripts] which might not have been written by a devout and liberal Jew, acquainted with Christian literature," his wife noted.[104] Boole probably never considered converting to Judaism, however. Surely he would have spent the rest of his life as a pariah, and Boole was already in a tight position as a Unitarian-leaning husband of a woman from a prominent British family. Boole would have to channel his impulse to return to a single, true faith elsewhere.

The Laws of Thought

It was Boole's formulation of symbolic logic, not any religious conversion, which satisfied this intense social and ideological longing for unity. Symbolic logic was based on a fundamental, noumenal language—mathematics—and it had the capacity to join disparate believers by providing an unbiased way to analyze contested doctrines. Creating such a mediating tool was a natural step for a teacher who preferred his students practice mock public meetings rather than cricket.[105] While writing the first draft of his groundbreaking essay, "The Mathematical Analysis of Logic" (1847), Boole paused to write poetry assailing "idle disputation," "wordy wrangling," "curious lore," and "folly born and ending in vexation."[106] In particular, religious thought and dialogue would have to be wrested from bickering, dogmatic clergymen. Boole impressed the greater significance of his technical work upon his wife, who later recounted that his guiding principle was "that theology always has been, and always will and must be, reformed from the outside, and very much from the side of science. He used to say that in any discussion among mere theologians, the worst side must almost necessarily win the day; inasmuch as theological discussions put a premium upon getting into an immoral and irreligious state of mind; and the man who wishes to be just and true hardly dares speak out, lest he should be cried down as lukewarm and unbelieving."[107] If faith and Christian brotherhood were to be rescued, humanity needed a method of arbitration, a way of calmly examining varying tenets and speculations. This was especially true in the tense religious climate of the 1840s and 1850s. Perhaps the most honest expression of George Boole's religious idealism, therefore, came in the form of his symbolic logic.

The publication of Boole's mature system, *An Investigation of the Laws of Thought, on Which Are Founded the Mathematical Theories of Logic and Probabilities* (London: Walton and Maberly, 1854), began a movement in philosophy toward algebraic symbols and away from abstract, ill-defined concepts. Although this shift is certainly of seismic proportions, it does not mean, however, that the book had an immediate impact. Boole himself realized that "it will be some time before its real nature is understood" and absorbed by a broad segment of readers.[108] One reason for the delay in comprehension is that *The Laws of Thought,* even a century and a half after its publication, remains a difficult read. Boole's prose is characteristically straightforward and crisp, but the book demands a wide range of expertise from its reader. A mathematician will find the long sections of algebraic proofs short on complexity, yet puzzle at Boole's digressions into "the constitution of the intellect"; a philosopher will catch Boole's sly deconstructions of Aristotle's *Metaphysics,* yet miss the subtle implications of his use of probability theory; a scholar of Victorian history will nod in understanding as Boole mentions contemporaries such as William Whewell, yet nod off as he spends more time analyzing the works of early modern philosophers such as Samuel Clarke and Spinoza using odd mathematical symbols and processes.

Furthermore, given Boole's religious idealism, his assertion that "it will be some time before its real nature is understood" was perhaps a double entendre—he may also have been hinting that he intended *The Laws of Thought* to be applicable to more than just the narrow field of logic. While editing the manuscript, Boole wrote a telling description of the project to Augustus De Morgan: "I do think that when we know all the scientific laws of the mind we shall be in a better position for a judgment on its metaphysical questions."[109] From the internal evidence of the text alone it undoubtedly aspired toward authority in many fields, scientific as well as religious, that Boole considered inextricably related. External evidence about Boole's life and beliefs only heightens such nontechnical concerns. As a friend of the Boole family wrote after studying *The Laws of Thought* for four years, it finally allowed her to understand the meaning of Hebrews 8:10: "This is the covenant that I will make with the house of Israel after those days, says the Lord: I will put my laws into their minds."[110]

Boole's ingenious method began with his boyhood revelation, nurtured by his Jewish mentor in Lincoln, that logicians could replace words with symbols and then manipulate those symbols—without care for what they represented—using some basic, rigorous rules. The first part of this insight was simple enough: we can represent any adjective or noun by algebraic letters. For instance, x could represent the set of "things that are black," or the adjective "black," and y could represent "dogs." After substituting symbols for words, mathematical laws suggest possible combinations and permutations. In our example, xy signifies the combination of "things that are black" and "dogs," that is, "black things that are also dogs" or "black dogs." Notably, the combination of x and y in the reverse order, yx, yields the same result: "dogs that are also black" or "black dogs." Hence Boole arrived at his first law of thought: $xy = yx$. The equation satisfied the mathematician: "The law expressed by $xy = yx$ may be characterized by saying that the literal symbols x, y, z, are *commutative, like the symbols of Algebra.*"[111] Boole was on his way to forming a new logic by using fundamental processes and rules from the grand discipline of mathematics.

Next, he realized that if two symbols mean the same thing their combination signifies nothing more than either one of them taken alone. If x stands for "dogs" and y stands for "canines," for instance, then xy, "dogs that are also canines," merely yields x (or y). In the form of an equation, $xy = x$ (or $xy = y$). Since x and y are synonymous ($x = y$), one can be substituted for the other. Replacing y by x in the equation $xy = x$ yields $xx = x$. Borrowing again from mathematics, Boole wrote this latter equation as $x^2 = x$—his second law of thought. He also proceeded to show how other basic rules of mathematics—including addition, subtraction, distribution, and transposition—also applied to his symbolic logic. Finally, to complete his system Boole added the logical concepts of "everything" (the "universe") and "nothing": 1 and 0, respectively.[112] The notation $1 - x$ thus provided a symbolic representation of all things not in the category described by x.

All of this perhaps looks and sounds simple enough, yet the mere transformation into an algebraic format led Boole to some penetrating insights. For example, within the second law of thought, $x^2 = x$, lies a relatively uncomplicated proof of one of the great laws of reasoning—the principle of contradiction, which states that an entity cannot both have and not have a given characteristic. Using subtraction, we can rewrite $x^2 = x$ as $x^2 - x = 0$, which in turn we can recast as $x(1 - x) = 0$ by using the law of distribution. Reading from left to right, $x(1 - x) = 0$ implies that the overlap between a group of things and everything not in that group is nothing. Let us say once again that x represents "dogs." Using Boole's notation, $1 - x$ therefore represents everything that is not a dog. The overlap between the group of everything that is not a dog and the group of dogs is clearly nothing. In other words, we cannot conceive of something that is at once a dog and not a dog. Through basic mathematics, Boole had arrived at a key principle that Aristotle took great pains to show in Book 4 of his *Metaphysics:* "'It is impossible for the same attribute at once to belong and not to belong to the same thing and in the same relation' . . . This is the most certain of all principles . . . [and] is by nature the source of all the other axioms."[113] For Boole, however, the principle of contradiction was not the foundation of other logical rules but merely a byproduct of $x^2 = x$.

Using his powerful new system, Boole went to work on contested doctrines. Although he employed secular illustrations for most of *The Laws of Thought,* religious cases crept into the latter half of the book. The first such example was an important passage on Kosher law: "Clean beasts are those which both divide the hoof and chew the cud." Boole broke this statement down into three parts, and converted those parts into algebraic notation: $x =$ clean beasts, $y =$ 'beasts with a divided hoof, and $z =$ beasts that chew cud. Reconstructing the original proposition using symbols gave him $x = yz$, and subtracting yz from each side of the equation produced $x - yz = 0$. Boole then squared the left side of the equation, since based on his second law of thought $(x^2 = x)$ this action should not matter. Using a common method of expanding squares in mathematics, Boole rewrote $(x - yz)^2 = 0$ in the unwieldy form $0xyz + xy(1 - z) + x(1 - y)z + x(1 - y)(1 - z) - (1 - x)yz + 0(1 - x)y(1 - z) + 0(1 - x)(1 - y)z + 0(1 - x)(1 - y)(1 - z) = 0$. Of course, since four terms on the left side equal 0 $[0xyz, 0(1 - x)y(1 - z), 0(1 - x)(1 - y)z,$ and $0(1 - x)(1 - y)(1 - z)]$, the formula could be simplified to $xy(1 - z) + x(1 - y)z + x(1 - y)(1 - z) - (1 - x)yz = 0$. Boole then equated each segment in the remaining left half of the equation to the right side of the equation, 0, to find his answer— that is, to discover which tenets the original proposition actually comprise. Boole's mathematical method thus yielded four new equations:

$$xy(1 - z) = 0 \qquad (1)$$
$$xz(1 - y) = 0 \qquad (2)$$
$$x(1 - y)(1 - z) = 0 \quad (3)$$
$$(1 - x)yz = 0 \qquad (4)$$

which signified that the original statement, "Clean beasts are those which both divide the hoof and chew the cud," entailed (again, reading left to right) the *nonexistence*

(1) Of beasts that are clean, and divide the hoof, but do not chew the cud.
(2) Of beasts that are clean, and chew the cud, but do not divide the hoof.
(3) Of beasts that are clean, and neither divide the hoof nor chew the cud.
(4) Of beasts that divide the hoof, and chew the cud, and are not clean.

To keep Kosher, therefore, Jews must maintain these four categorical denials.[114] The conclusions might appear obvious, and the mathematics unnecessary. Boole's underlying goal, however, was to show how people can calmly and precisely arrive at a full comprehension of the meaning of a doctrine—even a religious doctrine.

From this fairly basic example, Boole moved on to more complex analyses, culminating in an entire chapter on Spinoza and the early modern British philosopher Samuel Clarke. Boole's religious idealism helps to explain why he chose to deconstruct these particular thinkers as the centerpiece of *The Laws of Thought*. He believed that Clarke and Spinoza held a dangerous theological principle in common: that God, who is by definition an infinite Being and the single necessary entity of the universe, must exist "in every part of space."[115] In other words, the two philosophers inclined followers to look for the divine in the phenomenal, rather than noumenal, realm. Furthermore, Boole saw Clarke and Spinoza as masters of the "wordy wrangling" that obscured true faith: "The reasoning of Dr. Samuel Clarke is in part verbal, [and] that of Spinoza is so in a much greater degree; and perhaps this is the reason why, to some minds, [they have] appeared to possess a formal cogency."[116] Rhetorical agility combined with anti-idealist tendencies made the two early modern thinkers serious enemies of Boole and prime targets for his symbolic logic.

Using his mathematical system, Boole undermined the foundation of both philosophers' systems. His analysis of Clarke's proof of God provides the clearest example. Boole first quoted the theologian directly:

> For since something now is, 'tis manifest that something always was. Otherwise the things that now are must have risen out of nothing, absolutely and without cause. Which is a plain contraction in terms. For to say a thing is produced, and yet that there is no cause at all of that production, is to say that something is effected when it is effected by nothing, that is, at the same time when it is not effected at all. Whatever exists has a cause of its existence, either in the necessity of its own nature, and thus it must have been of itself eternal: or in the will of some other being, and then that other being must, at least in the order of nature and causality, have existed before it.[117]

Boole then reduced the convoluted passage into discrete statements:

x = Something is
y = Something always was
z = The things which now are have risen from nothing

p = It exists in the necessity of its own nature [i.e., the "something" spoken of above]

q = It exists by the will of another Being

and rephrased Clarke's argument symbolically, using v as an indefinite symbol:

(1) $x = 1$
(2) $x = v[y(1 - z) + z(1 - y)]$
(3) $x = v[p(1 - q) + q(1 - p)]$
(4) $p = vy$
(5) $q = v(1 - z)$

Eliminating v and arranging all five equations to equal zero gave Boole

(1) $1 - x = 0$
(2) $x[yz + (1 - y)(1 - z)] = 0$
(3) $x[pq + (1 - p)(1 - q)] = 0$
(4) $p(1 - y) = 0$
(5) $qz = 0$

Seeking to discover the truth or falsity of the original propositions y, z, p, and q now required the elimination of every other variable except the variable in question. Using the laws of substitution, addition, and subtraction, Boole boiled down equations (1) through (5) to $y = 1$ and $z = 0$, which indicated that "something always was" and "the things which are have not risen from nothing." However, trying to eliminate y to find p and q (the more important statements) proved problematic. Again using the five basic equations and mathematical laws, Boole arrived at $qz + pz = 0$. But since $z = 0$ this equation reduced further to $0 = 0$, which, as mathematicians know quite well, signifies absolutely nothing.[118] With obvious joy, Boole italicized his findings: "*Hence there is no conclusion derivable from the premises affirming the simple truth or falsehood of the proposition, 'The something which is exists in the necessity of its own nature'* . . . [and] *there is no conclusion deducible from the premises as to the simple truth or falsehood of the proposition, 'The something which is exists by the will of another Being.'*"[119] In short, Clarke's proof of the existence of God in space was meaningless. Under the bright light of Boole's symbolical method Spinoza fared equally poorly.

One might reasonably conclude from Boole's chapter on Clarke and Spinoza that he was an early logical positivist, exposing the absurdity of theological abstraction in the province of analytical philosophy. Boole, however, did not proceed in *The Laws of Thought* to undercut *all* discussions of religion through the use of mathematical logic—only those areas he believed were a source of pernicious confusion or distraction. He preempted the reader's concerns about this restriction of spiritual discourse with an admonishment at the beginning of his chapter on Spinoza and Clark: "The necessity of a rigorous determination of the real premises of a demonstration ought not to be regarded as an evil; especially as, when that task is accomplished, every source of doubt or ambiguity is removed." Boole further assuaged the fears of his audience by assuming the demeanor of a robotic arbitrator: "In employing the method of this treatise, the order in which premises are arranged, the mode of connexion which they exhibit, with every similar circumstance, may be esteemed a matter of indifference, and the process of inference is conducted with a precision which might almost be termed mechanical."[120] Boole therefore carefully laid his symbolic logic before the public as nothing more than a mediating tool, though one that might in the future "render service in the investigation of social problems."[121]

Of course, George Boole's real hope was that symbolic logic would lead to fewer sectarian conflicts and greater, more universal faith. The last chapter of *The Laws of Thought* tellingly assumed the form of a sermon, a strong caution to his readers. Deriding any notion of symbolic logic as mere ratiocination, Boole condemned "the too prevalent view of knowledge as merely secular" and worried that "in the extreme case it is not difficult to see that the continued operation of such motives, uncontrolled by any higher principles of action, uncorrected by the personal influence of superior minds, must tend to lower the standard of thought in reference to the objects of knowledge, and to render void and ineffectual whatsoever elements of a noble faith may still survive."[122] Should we fail to check those who deny the loftier realms of thought, Boole implied, we will be making the same mistake that Thomas Hill thought the ancient Romans made. Science stripped of its transcendental connections is uninspiring and debilitating, and a certain indicator of cultural decline. All conscientious intellectuals had to guard against this tremendous threat of secularization, Boole warned in the conclusion to his seminal work, unless they wished to find themselves in a society in which God has no place.

Throughout his life, he endeavored in the interest of the "noble faith" by engaging in wide-ranging attempts to solve religious disputes through the use of logic and mathematical notation. In unpublished (and undated) manuscripts, Boole closely examined religious doctrines from the entire history of the West. He tackled modern theology in a symbolical deconstruction of John Henry Newman's statements on the soul; he addressed ancient theology through a long and involved analysis of Plato's *Timaeus;* he even investigated theodicy, that prickly concern of nineteenth-century intellectuals, by manipulating equations derived from propositions about the origin of evil.[123] These passages show Boole grasping for the higher, divine plane where all knowledge converged into God's Truth.

Although many of his unpublished writings look similar to sections of *The Laws of Thought,* Boole apparently wrote some of these pieces before the completion of his mathematical system. For example, in an inchoate version of material he would put in his book, Boole substituted letters for statements from the Bible and then combined these symbols to reach conclusions. Using a passage from the Old Testament, Boole substituted the letter G for "They of faith are Abraham's sons," K for "men are either under the law or under faith," and L for "we are not under the law." He then argued that KLG confirmed "we are Abraham's sons."[124] The proof was fairly simple. KL means "we are under faith" because it is the combination of K, which sets up a binary option—we are either under faith or under the law—and L, which says that we are not under the law, leaving us therefore under faith. Subsequently, the product of K and L, "we are under faith," is combined with G, which asserts that all those under faith are sons of Abraham. Since we are under faith, and since those under faith are sons of Abraham, we faithful are, by extension, Abraham's sons. Boole performed the same type of rudimentary deductions with passages from Paul's Epistle to the Hebrews.[125]

Regardless of their complexity, such logical analyses were clearly motivated by, and in turn supported, Boole's religious idealism. Using what he considered to be the divine language of mathematics to comprehend a vast array of religious notions led him in a singular direction—the same direction as his father's telescope: upward toward heaven. "To infer the existence of an intelligent cause from the teeming evidences of surrounding design, to rise to the conception of a moral Governor of the world, from the study of the constitution and the moral provisions of our own nature . . . as these were the most ancient, so are they still the most solid foundations . . . of the belief that the course of this world is not abandoned to chance and inexorable fate," Boole concluded—"Revelation being set apart," of course.[126]

CHAPTER FOUR

Augustus De Morgan and
the Logic of Relations

An insignia designed by Augustus De Morgan for the nascent London Mathematical Society perhaps typifies the ecumenism of pure mathematicians in the Victorian age better than George Boole's poetry or Benjamin Peirce's prose. De Morgan inscribed a symmetric figure of concentric arcs and triangles with the name of the organization, the Latin motto *Vis unita fortior* (a united force is stronger), and three numbers: 1865, 5625, and 1281. The first number was probably the most obvious to contemporaries, being the year in which De Morgan, his son George, and a handful of other British mathematicians founded the institution. More precisely, 1865 was the year of the Society's founding based on the *Christian* calendar, for the other two dates were the year based on Jewish and Muslim reckoning. The motto amplified this interfaith overture, as De Morgan explicitly added it to highlight the "union of races and nations as well as of individuals" under the banner of mathematics.[1] Indeed, he further reinforced this idealistic unification through the symmetry of the figure, which De Morgan borrowed from the first chapter of the first book of Euclid's *Elements*. As he recorded in his notebook, this diagram represented the opening of the mathematical science to Jews, Christians, and Muslims alike.[2] De Morgan therefore felt he had created in the London Mathematical Society insignia an icon for the perfect and symbiotic whole of mathematics, humanity, and faith. This spirit lent one of De Morgan's technical phrases, "the logic of relations," a certain double entendre: it seemed logical to the mathematician to relate to people from all religious traditions.[3]

De Morgan's grittier side also gave his insignia for the London Mathematical Society a second, less spiritual meaning. The motto *Vis unita fortior* signified that mathematicians could enhance their authority by joining together in a professional organization such as the London Mathematical Society. In the early part of the nineteenth century, mathematical research comprised a loose federation of interested parties who approached somewhat nebulous topics by using a panoply of individual methods and symbols. By the end of the century, however, mathematics had become a far more centralized discipline, with numerous well-defined subjects of research and a standardized notation. Rather than often being a physicist, astronomer, or philosopher with a strong interest in mathematics, a "mathematician" was a distinct professional specializing in a circumscribed area. The London Mathematical Society was one of the first professional organizations specifically devoted to mathematics and helped to accelerate these changes. In the three decades after its founding, similar institutions spread across Europe and America. The sense of that impending professionalization of mathematical work is apparent in the London Mathematical Society insignia, uneasily coexisting with its loftier ecumenical sentiment.

To understand the professionalization of mathematics in the second half of the nineteenth century, we must examine how mathematicians portrayed their discipline in concert with their broader social concerns. In this context, the rise of modern, professional mathematics in large measure developed out of a strong desire among mathematicians like Augustus De Morgan to set their discipline apart from other spheres of Victorian society. "Gus! Gus! At 'em a'round!" was Augustus De Morgan's favorite anagram of his name, and he thought it a proper motto for his stand against the divisive nature of nineteenth-century religion, education, and intellectual life.[4] De Morgan protected and advanced the mathematical science by defining it, like the anagram, *against* what he saw around him. Where religious sects constantly bickered, mathematicians would discuss matters peacefully; where polemical fanatics overstated their cases, mathematicians would be cautious in their proclamations; where amateur mathematicians and arrogant metaphysicians discussed grand notions, professional mathematicians would limit their purview; where the evils of dogma and the religious establishment smothered nonconformity, mathematicians would be open to the new and different—as long as dogma and religion were not involved. In the setting of an extremely divisive culture, mathematicians would be active yet unified, progressive yet unassuming. Appealing to a conception of the "scientific method" that simultaneously embodied Pauline humility and professional ambition, mathematicians like De Morgan would distinguish themselves and their discipline.

To achieve this contradiction of Victorian culture and establish a professional realm for themselves, however, mathematicians would have to sacrifice the age-old transcendental characterization of their discipline. They could no longer claim that mathematics was a divine language because it then became a proper subject for clergymen and mystics as well; they could no longer assert that mathematics was perfect and infallible because it then became a new, dogmatic Church like the one they had struggled against; no longer could they even flaunt the supreme precision of mathematics because that was just the sort of hubris they disparaged in contemporary intellectual discourse. Strangely, therefore, to create a propitious environment for the professionalization of their discipline, practitioners such as De Morgan would have to treat their discipline with *less* respect, not more. After all, if mathematics was so potent why should it remain solely in the hands of parochial mathematicians? To advance their field, De Morgan and his colleagues would have to criticize it, circumscribe it—and, most importantly, secularize it—first.

A Briton Unattached

The most revealing feature of Augustus De Morgan's biography and personality was the way in which he considered himself free of the common identity markers of his age.[5] His father was a high-ranking official in the British command in India who was stationed during his career at numerous points across the subcontinent. In 1806, he transferred from Vellore just before one of the first mutinies by native troops. Augustus, the De Morgans' fifth child, was born later that year in the city of Madura, at the southern tip of India.[6] Although he spent only his first seven months in India before his family returned to England (a period surely lost to his memory), De Morgan used his expatriate birth to avoid allegiance to either a land or a sect. As he emphasized to his friend William Rowan Hamilton in 1852, "I am not *English*. I was born in India . . . in fact, I am a Briton unattached."[7] Similarly, he told his wife Sophia that if asked which religious denomination he belonged to, as the 1851 Census of Religious Worship demanded, he would write in "Christian unattached."[8] This lack of a clear denominational or cultural identity placed Augustus De Morgan, like George Boole in Ireland, in a position of alienation. The sense of being out of place, or of no place, was critical to the way De Morgan looked at the world. It spurred him to engage in teaching and research that he believed would surmount the divisions of the nineteenth century.

As a testy reformer with a deep antiauthoritarian streak, however, De Morgan advanced ecumenism far more intensely and publicly than his friend and cofounder of symbolic logic. Indeed, De Morgan's religious liberalism was part of a broad and often unsubtle progressivism that occasionally drifted toward sanctimony and a brusque demonization of his opponents. Following a spiritual crisis in the 1820s that separated the young man from his parents' evangelical faith, De Morgan fell in with prominent London radicals and nonconformists who shared his stridently antidogmatic and anticlerical position. He saw mathematics and his novel system of logic as an integral part of this radicalism, and (fairly or not) accused those he considered opponents of mathematics, such as the philosopher William Hamilton and the scientist Michael Faraday, with reproducing the errors of the religious establishment. In the same vein, De Morgan became one of the loudest advocates of research into spiritualism, less out of a wholesale belief in mediums than as a way to promote a healthy tolerance for the new and different.

De Morgan cultivated this independent spirit in reaction to an oppressive religious education. As a child, De Morgan's evangelical parents—especially his mother since his father had to return to India frequently during his childhood—force-fed him the Bible, inculcating an extreme distaste for biblical literalism. Young Augustus had to recite passages from Scripture ad nauseam and compose an abstract of each sermon he heard—no small task because the family went to church three times on Sundays.[9] Such overbearing parenting and emphasis on rote learning eventually resulted in spiritual disaffection in De Morgan's late adolescence. This ennui merely frightened his parents even further, of course. After he went off to Trinity College, Cambridge, in 1823, De Morgan's mother became worried that Augustus was going through the kind of religious transformation endemic among young intellectuals of the age. She sent a package of strictly orthodox books to him at college, enclosing a note that one can only imagine had an effect on the brooding teenager exactly opposite to the one his mother intended. "Can you wonder that a mother, doting as I do on you, feels miserable when she contemplates a beloved child wantonly sporting on the edge of so tremendous a precipice?" she wrote in response to De Morgan's drift away from the Bible and the Church of England. His mother darkly continued, "Can you picture to yourself any agonies like those which would take possession of your mind were you assured that before tomorrow morning you would be standing at the tremendous bar of an *angry God*[?]"[10] Maternal pleas, liberally sprinkled with terrifying eschatological threats, unsurprisingly failed to bring the young man into line with his parents' orthodoxy.

It was already too late: by the mid-1820s at Trinity, Augustus De Morgan was reading widely in religious literature without care for his parents' dogma. He was now in search of his own tenable spirituality. Most notably, he became particularly enamored by Berkeley and the idealism of that theologian's system.[11] Berkeley helped De Morgan through his crisis of faith by highlighting that true religion could exist outside of the realm of liturgy and man-made ecclesiastical structures. According to Sophia De Morgan, her husband's study of Berkeley led to an "absolute conviction, or, as he said, *consciousness,* of the fatherly care of God [which] was directly opposed to the scepticism" of influential contemporaries such as John Stuart Mill (who had come under the sway of Auguste Comte and Positivism).[12] Although De Morgan found Berkeley's religious philosophy less satisfying later in life, he nevertheless continued to pay respect by commenting on new editions of Berkeley's work and collecting first editions for his own library.[13] At the end of his time at Trinity, De Morgan was unable to take his master's degree because of an unwillingness to subscribe to the Thirty-nine Articles of the Church of England.[14] He later referred to subscriptions as "deadly poison" which "foster[ed] every kind of dishonesty" because in the realm of private conscience many of the faithful did not agree with the doctrines of their church.[15]

Following his years in Cambridge, De Morgan soon joined up with a circle of like-minded thinkers in and around University College, London (UCL).[16] In 1828, at the age of 21, De Morgan topped thirty-one other candidates to land a coveted position as the first professor of mathematics at the new university. UCL was to Augustus De Morgan as Queen's College, Cork, was to George Boole: a promising nonsectarian testing ground. From its inception it was also the epicenter of a great deal of progressive and dissenting thought. The idea for the university came from the Scottish poet Thomas Campbell, who, while traveling on the Continent, marveled at the new, religiously tolerant incarnation of the University of Bonn. Campbell found a wealthy backer for such a school in fellow Scot Henry Brougham, the Whig M.P. and founder of the *Edinburgh Review.* In London the two mustered additional support from high-profile liberals and nonconformists. Indeed, the founding committee of UCL read like a who's who of those excluded from the Oxbridge colleges: Isaac Lyon Goldsmid, later the first Jewish baronet; the Baptist minister Francis Augustus Cox; the abolitionist Zachary Macaulay; the Catholic Duke of Norfolk; and James Mill, the political theorist, disciple of Jeremy Bentham, and father of John Stuart Mill.[17] A self-proclaimed outsider, De Morgan could identify with all of these founders.

Established in 1826, UCL was from the start a school free of religious qualifications. Indeed, access for students of every sect and faith was guaranteed in the university's charter. Recorded (albeit inconspicuously) on the third page of the UCL prospectus was a declaration that the institution was "equally open to the youth of every religious persuasion," with religious education left to the students' parents.[18] The immediate influx of dissenters to UCL led to its derision within Anglican circles as "the Godless institution of Gower Street" and "the radical infidel College." Augustus Pugin called the neoclassical buildings of UCL "pagan" and snidely added that this was "in character with the intentions and principles of the institution."[19] The Tory press was the most aggressively derisive. One mocking ditty published by the "Privy Council of Stinkomalee" in the newspaper *John Bull* refrained, "Come, then, boys—my shirtless boys, who love such gay diversity,/No Church, no King, no '*nothing else*,' but Gow'r-street University."[20]

Augustus De Morgan, however, celebrated his participation in this bold academic experiment challenging the religious and civil establishments.[21] He hoped for the realization of Brougham's prediction that the new university would "do more to crush bigotry and intolerance than all the Bills [we] will ever see carried, at least until a Reform happens."[22] An early supporter of the Catholic Emancipation Bill, De Morgan enjoyed the chance to teach Roman Catholic students. In 1838, he even wrote to the chairman of the University of London executive council pressing for the admission of transfer students from a Roman Catholic college near Birmingham.[23] De Morgan also felt sympathy for his Jewish students, whom he saw as an unfairly "trounced community." The mathematician read Jewish newspapers (along with Boole), some of which forthrightly attacked Christianity and thus were anathema in England.[24] Many Jews, including the first M.A. graduate of the University of London, became lifelong friends of the De Morgans, who frequently entertained them at their home.[25]

Like Queen's College, Cork, UCL could not avoid controversies over authority and religion, and in due time it failed to become an idyllic sanctuary for the peaceful exchange of ideas among equals. Some devout professors, such as Thomas Dale, the first professor of English, felt uncomfortable with the school's evasion of religious matters and quickly moved on to other pursuits.[26] More seriously, lacking a well-defined administrative hierarchy and an explicit set of faculty rights, the university was wracked by power struggles soon after its creation. Particularly inflammatory were the actions of the college warden, Leonard Horner, who decided to expand his jurisdiction in the early spring of 1830. Horner declared he could make provisional orders in the place of the university council, including decisions about the fate of professors. An overly aggressive manager, he concentrated his efforts on removing Granville Sharp Pattison, the first professor of anatomy and the first faculty member to receive significant complaints about his skill and style.[27]

In a swift and sharp rebuke to Horner's activity in the Pattison case, De Morgan argued that the only power the college warden could rightfully claim was that "over the household servants."[28] For the remainder of the 1830–1831 school year the contentious 24-year-old professor pressed the trustees on the issue, worrying about the negative effects of placing too much authority in the hands of a single, opinionated individual.[29] On July 9, 1831, De Morgan raised the stakes by notifying the council that he would resign unless action was taken to ensure a latitude of opinion among the instructors.[30] The administration seemed intent on concentrating power in one way or another, though, to make the new school run smoothly. De Morgan in turn denounced this objective as amounting in practice to "the principle that a professor may be removed, and as far as you can do it, disgraced, without any fault of his own."[31] The mathematician finally gave up, handing in his resignation two weeks later.

The resignation did not sever De Morgan's ties to the liberal and radical ideologues who orbited UCL, however. In the 1830s, De Morgan wrote frequently for progressive outlets, receiving small payments for encyclopedia and journal articles. Some pieces were written anonymously; many covered mathematics and the history of mathematics. Most were published in adult-education compilations, marketed to middle-class aspirants who wished to increase their knowledge and gain a professional station in British society.[32] For example, De Morgan devoted enormous time to Brougham's Society for the Diffusion of Useful Knowledge (SDUK). Founded at the same time as UCL, the SDUK explained scientific concepts and techniques through plain-spoken pamphlets, monographs, and lectures.[33] Among other SDUK publications, De Morgan penned thirty-three articles for the *Quarterly Journal of Education* and a half-dozen for the *British Almanac*, as well as hundreds of smaller pieces for the *Penny Cyclopaedia*. Almost one-sixth of the latter, completed in 1858, was written by De Morgan.

De Morgan's involvement with the SDUK in his late twenties displayed his growing interest in promoting tolerance. His recollection of the organization's perspective is telling: "At its commencement the Society determined with obvious prudence to avoid the great subjects of religion and government, on which it was impossible to touch without provoking angry discussion." The members of the SDUK disliked how "religious disqualification and political exclusion occupied the daily attention of the press." Working-class violence, charged attempts at political reform, industrial upheaval, and previously unseen levels of religious dissension all formed the backdrop to the genesis and development of the organization. Like the founding members of the British Association for the Advancement of Science, the members of the SDUK sought to create a mutually beneficial union that would stand in marked contrast to the discord of the wider British culture of the 1830s.[34] "From the commencement the Society consisted of men of almost every religious persuasion. The harmony in which they have worked together is sufficient proof that there is nothing in difference of doctrinal creed which need prevent successful association when the object is good and the points of dispute are avoided," De Morgan happily remembered.[35] Envisioning the SDUK as an antisectarian body in constitution and outlook, he gladly became one of the group's most prolific contributors.

By his thirtieth birthday Augustus De Morgan had officially cast off the dogma of his parents. As Sophia De Morgan recalled about her husband's maturing view of Christianity, he now fervently held (much like Theodore Parker) that "the outward and visible sign can really stand for nothing when its inward and spiritual essence is gone."[36] "I conscientiously believe more than most of the clergy on whom you pin your faith," he proclaimed to his mother in 1836.[37] Regarding the New Testament, De Morgan failed to see how anyone could consider it supremely authoritative. Following the later sentiments of Coleridge and contemporaries who were affected by David Friedrich Strauss's *Leben Jesu,* including the theologian John Sterling and the historian and novelist James Anthony Froude, De Morgan challenged biblical literalism.[38] Moreover, he felt that the Church of England had distorted the Scriptures for its own intents and purposes, and called for a view of the Bible "unwarped by the devices of a Church of which it has always been the avowed doctrine to use every means which the age will allow to force men to agree to its own interpretations."[39]

De Morgan told his mother that a dispassionate, historical assessment of the New Testament revealed its malleability in the first centuries after the life of Jesus, and thus the text should be interpreted with the following saying in mind: "In the measure which you measure with, you shall be measured."[40] The Authorized Version, he emphasized, was no more than a translation of distant copies of fragments of contradictory accounts probably written by a multitude of authors. While many Christians are blissfully ignorant of this genealogy, De Morgan explained, "*I have been obliged* to consider this."[41] Such a consideration, he believed, laid bare the absurdity of dogma.[42] Admonishing against further interference, De Morgan delineated why no amount of intervention would convert him to her side:

1. Because I never saw or heard of any one who was made to change his [doctrinal] opinions by discussion.
2. Because such subjects are best discussed between a man and himself in retirement, and with the *real original accounts* before him.
3. Because I see in all that is orthodox a lack of that charity which Paul considers as more essential than everything else, coupled with what virtually amounts to a claim of infallibility.
4. Because numberless unanswered arguments lie before me, which the Established clergy have left off attempting to answer.[43]

The hubris of most orthodox believers clearly galled De Morgan, and his opposition to such overconfidence and inflexibility became a characteristic of his temperament that would resurface in other contexts later in life. With a flourish of profoundly contentious words, De Morgan told his mother that this would be his last exchange on the subject.

Ultimately, De Morgan decided that the question of biblical authorship was a matter for mathematical calculation, not polemical debate. Anticipating today's computerized analyses of texts, in which statistical profiles of word frequencies identify authors, De Morgan posited in 1851 that each author's prose had a specific mathematical fingerprint. One distinguishing number was the average length of words in a writer's compositions. Regarding this statistic, De Morgan asserted, "I should expect to find that one man writing on two different subjects agrees more nearly with himself than two different men writing on the same subject."[44] That is, since authors have a relatively constant style and vocabulary, a variety of texts by the same author should have similar word-length averages. For example, De Morgan estimated that different books in Thucydides's *History of the Peloponnesian*

War had word-length averages varying by no more than 15 thousandths of a letter—a very low statistical discrepancy—because he believed that 'Thucydides' was a single author.[45] In fact, given any text with a number of sections a mathematician could assess the likelihood of their common authorship through an objective measurement. With this mathematical theory and method in hand, scholars could solve the question of biblical authorship with relative ease. For instance, if researchers found significant variation in the word-length averages of two Pauline Epistles (again, in the mathematical sense, where a tenth of a letter is an enormous statistical variation in passages containing thousands of words), then it would be sensible to assume the existence of two different authors. Indeed, De Morgan wrote to William Rowan Hamilton in 1852 that he wished to use his technique to discover if Paul truly was the author of the Epistle to the Hebrews.[46] De Morgan hoped that such analyses, requiring only a modicum of mathematical skill, might someday be the primary line of scriptural inquiry.

Not coincidentally, Augustus De Morgan's abandonment of reverence for the Bible and his parents' faith occurred at the same time he began courting Sophia Frend (1809–1892), who harbored strongly unorthodox religious views of her own. Sophia followed in the footsteps of her father William Frend (1757–1841), a mathematician, linguist, and prominent nonconformist. Frend was teaching at Cambridge University in the late eighteenth century when he divulged and then aggressively promoted his Unitarian faith. After an ugly debate that in many ways prefigured the sectarian troubles encountered by De Morgan and Boole a generation later, Frend lost his teaching fellowship. He moved to London, where he established ties to other radicals, including Henry Brougham and Lady Byron. There he also met the young mathematician De Morgan, who enjoyed playing the flute at Frend's social parties. Thereafter William Frend functioned as a rebellious inspiration and ideological role model for De Morgan. De Morgan idealized Frend as "a splendid example of honesty in the pursuit of truth and of undaunted determination in the assertion of all that conscience required," and he hoped to emulate Frend's "chivalrous assertion of all that he thought true."[47] Indeed, for the short while until his death William Frend became the father figure Augustus De Morgan had often lacked in childhood. The two men maintained a warm relationship that included many lengthy days of heartfelt conversation.

De Morgan grudgingly had rejoined the faculty of UCL just before his marriage to Sophia Frend in 1837, but he subsequently kept a sharp eye on the college's dealings, especially on matters relating to religious sectarianism. For a decade and a half, as UCL underwent a period of great expansion in the 1840s and early 1850s, De Morgan found little to complain about. The same was not true, however, for his friends. In the early Victorian era, many of them—also ecumenical types, though generally more timid than De Morgan—found themselves under religious pressure at institutions of higher learning. George Boole's difficulties provided one of the most vivid examples. Throughout his life, De Morgan heard frequently from the co-founder of symbolic logic about the persistent troubles at Queen's College, Cork.[48] Although De Morgan had chosen an assertive public role, he appreciated Boole's hesitation to become directly involved in the sectarian disputes of the age. "I sincerely hope . . . that by keeping out of their squabbles, you may be able to live in peace," De Morgan wrote to his colleague in 1849.[49] Similarly, in the late 1840s one of De Morgan's friends at the Queen's College in Belfast found himself an outcast in a heated dispute over religion. The professor, J. R. Young, stayed on the sidelines during an ugly spat over the institution's funding and religious orientation, only to discover that his position had been erased in the upheaval. "Poor Young is no partizan," De Morgan lamented to William Rowan Hamilton in 1849, "and in the focus of religious dissention in which he lives such a person has no friends."[50] The experiences of Boole and Young demonstrated to De Morgan that sectarian fury had become a grave crisis.

After a respite, heated sectarianism returned to the UCL campus in 1853. In that year a wealthy benefactor left a significant amount of money to the college for the purchase of books, along with a problematic restriction: only Anglican professors could join the purchasing committee. The university library, which was building a collection from scratch, desperately wanted the money; in turn, the managing council felt inclined to accept it. Hearing of this inclination and bristling with indignation, De Morgan wrote to the council in blunt terms: "To my utter surprise, on the very first occasion on which money is offered on the condition of establishing a religious test, all I hear seems to indicate that it is far from certain that the offer will be rejected." Pressing the council to remain steadfast and reject the gift, De Morgan reminded the board of the animating principle of the college: "In the name of all the declarations which the College has put forth from its first institution, I claim the performance of the obligation therein undertaken to maintain every student, every Professor, every officer in perfect religious equality with the rest, from the President of the Council down to the sweeper of the floor."[51] In an adept display of bureaucratic guile, however, the council accepted the bequest, claiming that since the gift was not predicated on the hire or *normal* duties of any professor the religious qualification did not violate the college charter. The decision greatly angered De Morgan and he considered resigning again.[52]

It would take an affair of a far greater magnitude, though, to push De Morgan from UCL forever. Rather than quitting his professorship, for the next decade he would continue to use his position to advocate his deeply held belief that in an age of divisions one had to maintain an open mind and a robust sense of humility. From the late 1840s onward, De Morgan especially pursued this objective in the development of his new symbolic logic, in his investigation of spiritualism, and in his reconceptualization of the nature and scope of mathematics. Energetic thought and activity in these areas held the potential for social amelioration, De Morgan concluded, while also eroding the foundations of traditional authority and dogma.

Logic and Conflict

Around 1846 Augustus De Morgan became interested in logic, believing that the field had stagnated and was due for a reformation through modern mathematical techniques. As his work progressed toward remedying perceived deficiencies in the discipline, De Morgan wrote to contemporary philosophers to discuss logic's future. Never shy, De Morgan even sent a letter to William Hamilton, the prominent Scottish philosopher whom Thomas Hill had tried to outwit. Hamilton had appropriated the ideas of Immanuel Kant (as noted in chapter 2), in particular, the German philosopher's rigid separation of the phenomenal and noumenal realms, to give new life to the once great metaphysical school of Scotland. Besides philosophy, Hamilton also produced significant works in history, literature, physiology, and law. In short, the Scot was an intellectual giant who demanded the attention—positive or negative—of cerebral contemporaries.

When De Morgan sent Hamilton some thoughts on logic in the late 1840s, the philosopher was at the height of his power and influence, attracting a stable of dedicated students at the University of Edinburgh and receiving significant reviews in popular newspapers and journals. In response to De Morgan's queries, Hamilton sent the mathematician some comments and references to his own work on logic. Following their initial exchange, Hamilton—first in private letters, then quite publicly on the pages of the widely read weekly paper *The Athenaeum* and in a series of pamphlets—accused De Morgan of stealing his ideas about the quantification of the predicate. The accusation initially focused on the dates and content of their early correspondence, but soon expanded into broader areas of contention.

Simply put, in logic the predicate is what something is. In the sentence "This dog is black" the subject is "this dog" while the predicate is "black." In traditional Aristotelian logic the predicate could only have three states, or "quantities." A subject could be completely a predicate, not at all a predicate, or somewhat a predicate. In other words, a dog could be (completely) black, not at all black, or somewhat black. However, it became clear to many logicians in the nineteenth century that our observations take on many more forms than Aristotelian logic allows. In fact, the previous sentence is a good example: nineteenth-century logicians envisioned *more* forms than Aristotle did. Adding quantifications of the predicate such as "more" and "less" was absolutely critical to the progress of logic. With only "all," "none," and "some," the latitude of the syllogism was greatly restricted.

Being able to understand statements like "This dog is blacker than that dog" or "Some dogs are some pets" occupied De Morgan and Hamilton beginning in the 1840s, eventually leading to an emotional, and highly revealing, feud over a seemingly abstruse topic. Looking back from a century and a half later, Hamilton's anger seems grossly unjustified; his logic took shape in a way that barely overlapped with De Morgan's symbolical forms. At the time, however, the debate was one of the longest, most intense, and most celebrated philosophical controversies of the age. The dispute over whether the quantification of the predicate had a philosophical or mathematical origin even outlived Hamilton's death in 1859, as Hamiltonian philosophers defended their patron saint against the small but growing ranks of mathematical logicians.[53]

Why such intensity? The terms of the debate and its portrayal by each side show that the argument was less about priority and plagiarism than about William Hamilton's and Augustus De Morgan's contrasting senses of mathematics, metaphysics, and authority. Hamilton, whom De Morgan referred to during their indelicate *pas de deux* as "The Prince of Darkness,"[54] was the epitome of an antimathematical philosopher. Defending his claim as the progenitor of new logical forms became an occasion for Hamilton to castigate mathematicians like Boole and De Morgan for entering into what had been the exclusive domain of the philosopher for over two millennia.[55]

Hamilton's lectures on logic maintained a strict tone against mathematical interlopers while upholding the superiority of his traditional linguistic method.

In more widely accessible works such as his *Discussions on Philosophy and Literature* (published in 1852, but consisting of articles from the prior two decades), the philosopher bluntly indicated his dislike of mathematics. Attacking William Whewell's treatise on the value of mathematics in liberal education,[56] Hamilton judged mathematics as grossly inferior to philosophy. "If we consult reason, experience, and the common testimony of ancient and modern times," Hamilton wrote, "none of our intellectual studies tend to cultivate *a smaller number of the faculties, in a more partial or feeble manner, than mathematics.*"[57] Adopting a scholastic tone destined to irk De Morgan even further, Hamilton quoted from numerous "authorities" who allegedly supported this characterization of mathematics.[58] Dropping all pretense of cordiality, the philosopher concluded that mathematics was a pox on the educational system: "An excessive study of the mathematical sciences not only does not prepare, but absolutely *incapacitates the mind,* for those intellectual energies which philosophy and life require."[59] Rubbing salt in the wound, Hamilton also declared that mathematics was too simple for great minds and thus made its practitioners unbecomingly dull.[60] It was for good reason, therefore, that De Morgan referred to Hamilton as the "assailant of mathematics."[61] The Scot labeled the discipline as an extremely low, rudimentary form of thinking that lay far beneath the stratospheric realm of the metaphysician.

On the other side of the debate, De Morgan associated Hamilton with dangerous ideas and attitudes that he believed threatened the Victorian age with religious distraction and dogmatism. Here De Morgan found common ground with John Stuart Mill, whom the mathematician previously had derided for his foray into Comtean Positivism but who now appeared as a staunch, welcome antagonist of the Scottish metaphysician. Like Mill, De Morgan thought that Hamilton's discipline had a misguided sense of priorities, overvaluing certainty in a way that guaranteed never-ending disputes. "There are no writers who give us so much *must* with so little *why,* as the metaphysicians," De Morgan remarked disparagingly to the English mathematician and astronomer John Herschel regarding the state of philosophy in the 1840s. Audacity combined with a failure to justify assumptions made these metaphysicians akin to the sectarian zealots De Morgan repudiated in his public declarations and social activities. "He who said *peace among men* forbade Metaphysics," De Morgan told his friend.[62] When William Hamilton returned a gratis copy of De Morgan's *Formal Logic* (1847), the mathematician summarized his negative view of metaphysics in the margins of the book. A grand philosopher like Hamilton was essentially incontrovertible, since others have "no means of verifying his conclusions." Inverting (perhaps unconsciously) Plato's famous metaphor of the cave, De Morgan described metaphysicians as foolishly burrowing into the ground rather than bravely struggling out into the sunlight of truth:

> The metaphysician is down at the granite, with an eternal effort to see what is under it. After breaking a dozen pickaxes, and tiring his bones, he finds it easier to throw down his tools, and to think his way to the centre; and as nobody can contradict him with a pickaxe, which is the sole comfort of his position, he can declare what he imagines himself to have arrived at with any degree of dogmatism which his conscience will allow him to use. . . . Nothing is so easy as certainty: and nothing gives such a certainty of certainty as the study of an area in which the results cannot be tested.[63]

In mathematical terms, metaphysics had no Archimedean point—and thus, perhaps, no point at all but to stir incessant controversy.[64]

Philosophical idealism, which Hamilton incorporated into his thought, seemed a particularly egregious offender in this respect. Although he had flirted with idealism in his youth, notably in his attraction to Berkeley, in the 1840s De Morgan moved swiftly away from such ethereal notions as he became increasingly hardheaded and pragmatic. Grand metaphysical interpretations and discussions of the ideal realm quickly lost their appeal. "To read Berkeley so as to give him a fair chance, some one else should turn the page over," De Morgan teased in 1847, "for, unphilosophical as it may be, the touch of the paper periodically intervening is a snake in the grass—an unphilosophical snake. It is hard to make philosophers of the fingers."[65] Hamilton's idealism came out of more recent German thought, not Berkeley, of course; yet this merely placed De Morgan at greater odds with his antagonist's system. As the mathematician disapprovingly wrote in the 1850s, Hamilton's central notion of the "Unconditioned" appeared to lead to "two German equations—Everything = God. God = 0"—that is, a pantheism bordering on atheism.[66] De Morgan may have been religiously liberal, but he found the vogue for such vague concepts unpalatable. Indeed, he spent over a decade trying to debunk the "Unconditioned," convinced that the allure of the term was due to its extreme (and perhaps intentional) nebulosity, much like the countless sectarian doctrines his contemporaries uncritically accepted.[67]

De Morgan thus portrayed Hamilton and his fellow philosophers as part of an old guard, like the clergy—remnants of an age in which obscurantism and authority ruled. In an 1852 letter to William Rowan Hamilton, the mathematician, De Morgan conceived the philosopher's operating principle to be "[t]hat any man, having a point left incomplete, may make it up by ancient authority. If you boggle at [this] postulate, I refer you to your namesake[68] at Edinburgh, and his followers, who never hesitate at imposing a law of thought as necessary, if Aristotle has declared it. I am not joking. If you or I can see no necessity they refer you to Aristotle to prove it."[69] Moreover, as the nineteenth century progressed it seemed to De Morgan that William Hamilton had been elevated, like Aristotle, into an untouchable philosophical pantheon. A cult had arisen around the Scot and his thought, including a tenacious set of disciples and a host of intellectuals who analyzed his works with the intensity of biblical exegesis.

Appropriating logic from metaphysicians such as Hamilton was more than a technical necessity, De Morgan believed—it went to the heart of his age's cultural problems. The consequences of philosophy's stranglehold on logic were grave, since without a rigorous system, philosophers, theologians, and others could advance rickety or deceptive theories with impunity. As De Morgan ruefully noted to a friend: "The world does not know the amount of material mischief done by want of logic. . . . an immense quantity of power is deducted by this."[70] He thus considered himself at war over the ownership of a crucial field, one that could reduce dogmatism and societal tensions if only it were properly structured.

Indeed, De Morgan believed his new logical system would provide a dispassionate technique for human beings to discuss and analyze their ideas in a reasonable, considerate, and open way. A technically exact symbolic logic would be a bulwark against partisanship, he wrote at the beginning of *Formal Logic*: "The confidence which the favourers of . . . several theories place in their correctness is a sufficient reason for keeping the account of the process of the understanding, so far as it can be made an exact science, as distinct as possible from all of them: for they differ widely, and if they agree in anything which can be distinctly apprehended, it is only in having names of great authority enrolled among the partisans of every one."[71] Upon closer inspection, seemingly unimpeachable support for rigid opinions frequently came merely from a cursory reference to famous thinkers and a style of reasoning peculiar to the specific viewpoint in question. By having all parties agree to the same method of logical proof, these illusory forms of support would rapidly fade and the opportunity to build common ground would greatly increase.

In *Formal Logic*, published the same week as George Boole's preliminary effort "The Mathematical Analysis of Logic," De Morgan's social agenda and professional research have an unexpected, yet thoroughly natural, affinity. For a supposedly technical treatise, the book reads as much like a primer on the human propensity to be dogmatic as it does an intricate analysis of logical properties. One of the monograph's central theses is that (sadly) human beings are deeply inclined to have steadfast opinions, even on matters they know nothing about. A reasonable, moderate disposition is quite rare, De Morgan observed: "Many minds, and almost all uneducated ones, can hardly retain an intermediate state" regarding their beliefs. Nearly any query would prove this point: "Put it to the first comer, what he thinks on the question whether there be volcanoes on the unseen side of the moon larger than those on our side. The odds are, that though he has never thought of the question, he has a pretty stiff opinion in three seconds."[72]

In one particularly detailed section of *Formal Logic*, De Morgan enumerated the rhetorical tactics of overconfident partisans.[73] "Controversialists constantly lay too much stress on their own negative proofs [of their opponents' arguments], on their *I cannot find*, even as to cases in which it is palpably not their interest to find," one example highlighted.[74] While myopic extremists constantly understated and thus undermined their rivals' positions, they in turn overstated the strength of their own positions. De Morgan complained that controversialists "must decide for [the reader], and thus act both counsel and judge: probably because their arguments are not so convincing to their own minds as they wish them to be to the reader's. They prove, at the utmost, their own conviction that they have the right side: but the thing to be proved is that such conviction is well founded."[75] Zealots construct a spurious certitude out of slight evidence, leading to widespread—and entirely unjustified—antagonism.[76] In short, De Morgan accused those who believed they held the sole way of comprehending the world of the sin of hubris, and with a heavy ethical tone he inveighed against such ubiquitous, misguided fanaticism throughout *Formal Logic*.

Furthermore, the vague application of language greatly intensified this partisanship. "The growth of inaccurate expression" arising from a poor understanding of language and logic, De Morgan proclaimed, "gives us swarms of legislators, preachers, and teachers of all kinds, who can only deal with their own meaning as bad spellers deal with a hard word, put together letters which give a certain resemblance, more or less as the case may be."[77] Echoing the third book of John Locke's *Essay Concerning Human Understanding*, De Morgan argued that a lack of clarity in language leads to confusion and even outright and willful abuse by unscrupulous parties.[78] Moreover, he felt this "ambiguity in [the] meaning of words" exacerbated his era's animosity. Locke's abstract linguistic problem had become a very real and widespread social dilemma.[79]

Although De Morgan thought Locke had properly analyzed the problems of language and communication, his solution—an appeal for toleration—appeared a bit too idealistic. De Morgan found himself in an age of proliferating, competing sects—precisely what Locke had yearned for—yet the nineteenth-century mathematician (as well as many other Victorian intellectuals) considered this world in many ways inferior to the stability of medieval Britain. Locke's notion that each individual could find his own path to God if freed from the fetters of government and the Established Church neglected to account for the multifaceted discord engendered by a liberal society. Endless friction between sects—as profoundly unsettling as the military conflicts of the seventeenth century—might well be the result of religious freedom. By contrast, to effect a more cohesive civil society, De Morgan envisioned his project as not only reining in the authorities that worried Locke, but also as weakening the dogmatic bias and wordplay of *all* partisans.

To produce this peaceful coexistence, De Morgan sought a method for making declarations and arguments as precise as algebra or geometry. Unlike regular speech—and certainly unlike common theological and philosophical discourse—mathematics maintained a highly instructive clarity and exactness, De Morgan believed. Mathematics also did not rely on scanty evidence, nebulous terms, and tenuously made arguments. De Morgan strove to emphasize these points to his students. As opposed to everyday speech, mathematics has "no terms exactly synonymous, and those between which there is most resemblance, can be as clearly distinguished from one another as others which have the most contrary significations," he wrote in one of his very first lectures.[80]

The distinctive, recondite nature of De Morgan's symbolical system of logic reflected his urge to emulate pure mathematics in the sphere of language. Like Boole, De Morgan found certain mathematical symbols irresistible. His first appropriations were the fundamental signs "+," "−," and "=," which (with greater exactitude) took the place of the basic notions "and," "not," and "is." He also discovered that algebraic variables (such as X and Y) could be natural replacements for logical categories. Rather than adding new, complex symbols to these rudimentary signifiers, De Morgan pragmatically added two common symbols that would facilitate widespread use of his system in print: parentheses and periods.

Placed next to variables, these two symbols were the key to his logical notation. Using them, De Morgan carefully distilled all possible propositions into eight specific forms. The first six rephrased Aristotelian logic: "X))Y" meant that everything in the category X is also in the category Y; similarly, "X((Y" signified that every Y is X; "X(.(Y" signified that some Xs are not Ys, while "X).)Y" signified that some Ys are not Xs; "X).(Y" signified that no X is Y, while "X()Y" meant that some Xs are Ys. Just as Boole's use of mathematical notation suggested additional logical forms and insights that eluded Aristotle, so did De Morgan's symbolism. To the six Aristotelian forms De Morgan added "X(.)Y," signifying that everything is X or Y or both, and "X)(Y," signifying that some things are neither Xs nor Ys.[81] With this new notation and an enlarged set of possible propositions, De Morgan could extend and refine logic's potential. Scholars would be able to analyze a greater variety of statements, and they could examine those statements with a level of subtlety—and exactitude—unavailable in the two millennia since Aristotle.

Like George Boole, Augustus De Morgan thought this revolutionary precision would be especially useful in deterring or preventing antagonists from polarizing religious debates. Reflecting on a favorite quip about the Bible, "One day at least in every week/The sects of every kind/Their doctrines here are sure to seek/And just as sure to find," De Morgan saw the problem's roots in the confusion of language.[82] Indeed, following the development of his symbolic logic De Morgan ceaselessly attacked the haziness of the theological terms used in his age. For instance, in an 1852 letter discussing the various shades of meaning in the Athanasian and Arian creeds, De Morgan threw up his hands: "Primary and secondary meanings of words are so mingled, that two persons who communicate on the subject [of Christian theology] need a *dictionary* of definitions—not a *chapter*."[83]

Even the basic religious terms "belief," "worship," and "subject" had evolving meanings over the centuries, clouding the proper interpretation of countless religious tenets.[84] More recently, the mathematician believed, people had carelessly oversimplified debates by using mutually exclusive words like "churchman" and "heterodox."[85] De Morgan's public speculations about the etymology of such contested Christian words as "hypostasis" and "substance" earned him nasty partisan rebukes, particularly from those who knew of his Unitarian bent.[86] Tellingly, late in life De Morgan spent time comparing the meanings of key phrases in different versions of the Greek Scriptures.[87]

Such concerns underscore how De Morgan's creation of a symbolic logic was a religious and social act as well as a technical achievement. A precise understanding of terms and the relationships between them would allow for a more nuanced and thus less contentious dialogue. Black-and-white portrayals of issues would give way to a spectrum of thought that better represented the feelings of well-intentioned people. Support for every conclusion would have to come through logical methods accepted even by those who were disinclined to the position in question. Rigidity would give way to mutual appreciation. De Morgan understood, however, that this process would not occur overnight. When he reiterated the need for a "great deal" of additional research into logic in his presidential address to the first meeting of the London Mathematical Society in 1865—almost two decades after the publication of *Formal Logic*—the audience sensed a greater mission than merely the expansion of his symbolical method.[88] As De Morgan cautioned his fellow mathematicians, "If we do not attend to extension of language, we are shut in and confined by it."[89]

De Morgan and Spiritualism

Augustus De Morgan's interest in, and respect for, psychics naturally complemented his attacks on unchallenged dogma and inflexible zealotry. Janet Oppenheim briefly noted this attraction in *The Other World: Spiritualism and Psychical Research in England, 1850–1914*, placing De Morgan in the context of numerous contemporary intellectuals who shared such enthusiasm.[90] Oppenheim could have written considerably more on the mathematician; he spent much of his later life investigating mediums and spiritualist activity. Along with his fascination with mysterious occurrences, spiritualism posed a litmus test: to commit to psychical research and the *possibility* of spiritualism meant accepting an uncertain state of knowledge, a temper De Morgan thought appropriate for the time.

Published and unpublished sources show that De Morgan, in concert with his wife Sophia, became attracted to the occult arts during the 1840s—following his break from his parent's faith, and at about the same time he began work on his symbolic logic. Writing to William Rowan Hamilton in 1847, De Morgan described his tentative investigations of psychic power: "I am myself past thinking anything too extraordinary to be true . . . We know that every motion of every particle of matter has its effect upon the motion of every other particle . . . Now is there a *mental dynamics?*" He also pondered why ghosts have been observed in every culture, a puzzling ubiquity deserving at least some exploration.[91] De Morgan went on to examine various paranormal affairs in the ensuing years. By the end of the 1840s he was well briefed in mesmerism and clairvoyance, and wondered aloud how mere coincidence could explain mind reading.[92] As De Morgan frequently emphasized to his friends, his impaired vision made him inherently suspicious of any *trompe l'oeil*—"As to my eyes—or eye—I don't believe them (it) much—never had much reason" he once wrote—which made his acceptance of the strange incidents he witnessed carry that much more weight.[93]

In the early 1850s the De Morgans advanced their interest in spiritualism by attending, and ultimately hosting, seances. The summoning of mysterious rapping noises that began with the Fox sisters of Hydesville, New York, in 1848, spread rapidly overseas in these years. Superstar mediums traveled from the United States to Britain, touching off astonishment, research, and more than a bit of furor.[94] Two of these American mediums had an especially profound impact on the De Morgans: Daniel D. Home and Mrs. W. R. Hayden. Hayden was the more restrained and subtle of the two—she published no autohagiographies, as the flamboyant Home did—yet she was the first to travel abroad and her powers were as well regarded as her male counterpart's. In October 1852, Hayden arrived on British soil to a groundswell of interest among the educated upper-middle class.[95] A few months later, the De Morgans excitedly invited Hayden to their house in London so they could personally interrogate the spirits she claimed to transmit.

Hayden clearly stunned the De Morgans and their half-dozen guests. Indeed, the experience had such a visceral impact on Augustus and Sophia that they would recount details of the night for the next decade, in publications and testimony for others interested in spiritualism.[96] Hayden appeared to summon multiple spirits on command, producing a distinct rapping sound for each one while the guests peeked under the table to ensure she was not using her legs. Augustus was astounded when the rapping spelled out letters relating to an obscure article he had read many years earlier.[97] "I have been hearing the spirits, or whatever they are, rap upon the table, and answer questions put *mentally* . . . This is more puzzling than $\sqrt{-1}$. . . . Our *thoughts are read*, somehow, you may swear," he ecstatically wrote to William Rowan Hamilton afterward.[98]

In a moment akin to spiritual revelation, the spellbinding Mrs. Hayden solidified Augustus De Morgan's belief that something extraordinary occurred during seances. Although he could not commit himself wholeheartedly to spiritualism, no longer could De Morgan accept scientific dismissals of the phenomena. "If I were bound to choose among things which I can conceive," he later summarized in the preface to his wife's compendium of their spiritualist encounters, *From Matter to Spirit* (1863),[99] "I should say that there is some sort of action of some combination of will, intellect, and physical power, which is not that of any of the human beings present."[100] For De Morgan, these immediate and visceral seances hinted at a spiritual realm that was palpable and responsive as opposed to metaphysical and distanced from humanity (as his nemesis William Hamilton held). With this conviction, De Morgan became one of the greatest Victorian advocates of psychical research.

His profile was quite typical for a nineteenth-century intellectual strongly attracted to spiritualism: someone who had cast off biblical literalism and the Church of England but who still had strong religious yearnings and wished to pursue those feelings in an ecumenical way.[101] In correspondence, he clearly relished the credence given to spiritualism and mediums by people from across the religious spectrum: Protestants, Roman Catholics, Swedenborgians, and Millennialists alike.[102] Moreover, he took the opportunity during the rise of spiritualism in the 1850s and 1860s to correspond with a wide variety of believers. He exchanged notes on mesmerism, clairvoyance, and telepathy with Samuel Maitland, a Roman Catholic priest who wrote on spiritualism and miracles,[103] as well as Alfred Russel Wallace, the naturalist cofounder of evolutionary theory.[104] Furthermore, De Morgan reveled in how an increasing number of like-minded progressives were coming to believe in the existence of spirits. When another American medium arrived in London in the summer of 1855 to flourishing interest, De Morgan followed the story with glee as the medium "reedited all the laws of nature" before many of his credulous, liberal friends.[105]

The overlap between spiritualist and radical circles, apparent from the start, intensified during the 1860s,[106] and correspondingly, Augustus De Morgan linked antispiritualism with political, religious, and institutional conservatism. For De Morgan the man who most embodied these associations was Michael Faraday, the pioneer of electromagnetism and prime mover of the Royal Institution. Although Faraday investigated extraordinary, invisible forces himself (and even may have conceived the unity of these forces for mystical reasons),[107] he was one of the most prominent and outspoken antispiritualists of the day. As Faraday wrote to a friend while Mrs. Hayden traveled around Britain, he dreamt of "turning the tables upon the table-turners."[108] In both *The Times* and *The Athenaeum,* the scientist labeled spiritualists as crackpots and castigated those who had fallen under the mediums' spell.[109] Faraday even ran a series of experiments to debunk the mediums, which he believed were conclusive but which did little to convince those fascinated by seances, including De Morgan.[110]

De Morgan predictably took issue with Faraday's headstrong antispiritualism for over a decade, but, as with his feud with William Hamilton, the intellectual conflict had significant ulterior motives. Faraday was devoted to Sandemanianism, the Christian sect founded by the outcast Scottish minister John Glas (1695–1773), which flourished in Britain's age of industrialization. In broad terms, Glas could be called an early fundamentalist, emphasizing a strict literal reading of the Bible and opposing the value of political reform. Although Glas's opposition to the Established Church might have won the applause of De Morgan, the mathematician was repelled by Sandemanianism's overriding focus on the Bible and its belief that no room existed for varying opinions on religious matters. As Faraday's biographer Geoffrey Cantor has succinctly put it, "The Sandemanians sought to understand every word in the Bible without introducing distortions of human origin."[111] Of course, after the 1820s De Morgan viewed the Bible as nothing *but* distortion and held that "God intended religion as a means of developing the inquisitive faculties."[112] Michael Faraday's attachment to Sandemanianism thus meant that despite his scientific endeavors he took the Bible literally and maintained a conservative political outlook—in other words, he was diametrically opposed to De Morgan's perspectives on these matters. De Morgan's defense of psychical research against Faraday therefore had a broad sentiment behind it. De Morgan associated Faraday's unwillingness to allow for the possibility of spiritualism with his inflexibility in religious and political matters.

To make their relationship even more discordant, Faraday was also, like William Hamilton, arrogantly (and, for a scientist, perhaps oddly) anti-mathematical. Self-educated and predisposed to hands-on experimental discoveries, Faraday admitted he was ignorant of advanced mathematics, which prevented him from understanding much of his colleagues' work.[113] Moreover, he seems to have had a deep-seated hatred of the discipline. Faraday relentlessly attacked the value of pure mathematics in the school curriculum, advocating studies of technology and the applied sciences instead.[114] As his successor at the Royal Institution, John Tyndall, wrote, "No man felt this tyranny of symbols more deeply than Faraday, and no man was ever more assiduous than he to liberate himself from them, and the terms which suggested them."[115] This phobia almost certainly had a religious root: Sandemanians were opposed to man-made signs and languages, which they believed were profane and unfit to supersede the signs of the Bible.[116]

Faraday's antimathematical bent greatly heightened De Morgan's hostility toward him, making their exchanges unusually caustic. For instance, in an 1854 lecture ostensibly on "Mental Education" and attended by numerous luminaries including Prince Albert, Faraday took the opportunity to criticize spiritualists and psychical researchers like De Morgan for utterly lacking sound judgment. "Before we proceed to consider any questions involving physical principles," Faraday told the audience, "we should set out with clear ideas of the naturally possible and impossible."[117] In response, De Morgan charged that scientific organizations such as Faraday's Royal Institution had become too conservative. "The scientific bodies are far too well established to risk themselves," he wrote in an extremely harsh review of Faraday's lecture, "These great institutions are now without any collective purpose except that of promoting individual energy: they print for their contributors."[118] Rebuking Faraday—while implicitly defending psychical research—De Morgan concluded, "We thought that mature minds were rather inclined to believe that a knowledge of the limits of possibility and impossibility was only the mirage which constantly recedes as we approach it."[119] Here, the mathematician's visions of religion and science clearly overlapped. Religious dogma and rigid scientific assumptions appeared to De Morgan as part of the same unseemly attitude. As he had written during the turmoil of his early twenties, when he had refused to subscribe to the Thirty-nine Articles of the Church of England, no error damaged thought and morals more than the promulgation of "self-evident" doctrines—particularly by those in positions of authority.[120]

In turn De Morgan characterized those interested in spiritualism as appropriately humble and broad-minded, which he considered the true character of science from Newton onward (as well as the proper religious demeanor). Psychical researchers "have the *spirit* and the *method* of the grand time when those paths were cut through the uncleared forest in which it is now the daily routine to walk," De Morgan claimed in *From Matter to Spirit*. He continued, with emotion, "What was that spirit? It was the spirit of universal examination, wholly unchecked by fear of being detected in the investigation of nonsense."[121] In a direct affront to Faraday's strong conviction that psychical researchers were pseudo- or even antiscientific, De Morgan asserted that they were actually on the cutting edge of research. He painted Faraday and others like him as antiliberal curmudgeons: "The Spiritualists, beyond a doubt, are in the track that has led to all advancement in physical science: their opponents are the representatives of those who have striven against progress."[122]

Much of De Morgan's writing on spiritualism sounds conspicuously similar to his work on logic—full of passages assailing dogma, presumption, and authority. Generalizing from the case of Faraday, De Morgan had a derogatory name for all those who arrogantly defined the boundaries of investigation and belief, thus hindering the pursuit of truth and the progress of mankind. He called them "philosophers," and he demonized them as no better than the dogmatic clergy: "There is philosophercraft as well as priestcraft, both from one source, both of one spirit."[123] As De Morgan critically observed in *From Matter to Spirit*, "The commonest of all questions is, 'How do you account for . . . ?' and woe to him who, not having an answer of his own, shall refuse to accept that of the querist."[124] Moreover, the mathematician saw his contemporaries as "so nourished on theories, hypotheses, and other things to be desired to make us wise, that most of us cannot live with an unexplained fact in our heads. If we knew that omniscience would reveal the secret in a quarter of an hour, we should in one minute have contrived something on which to last through the other fourteen."[125] The culture of the Victorian era demonstrated how opinionated and close-minded people could be. "Try to balance a level on the palm of the hand with the bubble in the middle," De Morgan metaphorically asked, "Who can do it? Not one in a hundred. The little air-drop is always in extremes: it may stay in the interval for a few seconds, and then comes a tiny unconscious motion which sends it right up to one end or the other."[126] He yearned for this elusive balance, an attitude of modesty and reserved judgment that he increasingly identified as the most propitious disposition for his age.

The Temperament of Professional Mathematics

Augustus De Morgan ultimately envisioned professional mathematics as the foremost representative of this humble disposition. He set his discipline against the division and rigidity of sectarian faiths, the authoritarian realm of philosophy, and the conservatism of institutional science. Beginning in the 1840s, De Morgan's view of the scope and subject matter of mathematics reflected this need to develop a temperament directly counter to the disposition of William Hamilton, Michael Faraday, and many clergymen of the day.

De Morgan's fight with Hamilton in the late 1840s and early 1850s clearly strengthened his desire to curtail often divisive academic issues, such as debates over credit for a discovery. As De Morgan worriedly wrote to a friend at the beginning of 1852, "The mathematicians at home and abroad are getting into a somewhat fidgety and excitable state about priority. For a great many years I have noticed a somewhat augmenting tendency to guard themselves against others or others against themselves."[127] As mathematics began to undergo serious professionalization in the middle of the nineteenth century, claims of precedence indeed rose sharply. With researchers arguing about credit for each minute theoretical advance, a great deal of productive energy was assuredly lost, De Morgan believed. The egotistical disputes clearly enervated him. "As to discussions on priority questions I am too wary . . . what a sad thing it is that mathematicians should fall out, and chide, and fight," he mourned in early 1853.[128] At other times, De Morgan simply had to laugh at how ridiculous the controversies seemed, such as when numerous mathematicians claimed they had discovered $\sqrt{-1}$, even though no one knew exactly what it meant yet.[129] Apparently many theorists merely wished to achieve fame rather than advance their discipline. The absurdity of the backbiting seemed obvious to De Morgan, as he wisecracked in a letter written near midnight: "So no more at present, for if I write five minutes more, I shall be obliged to alter the date, and since the additional verse is *discovery,* I may lose the priority thereby, so fast do people jump on each other's heels now-a-days."[130] In 1854, De Morgan asked for William Rowan Hamilton's help in admonishing young mathematicians about the evils of being concerned with priority.[131]

A historian of mathematics as well as a practitioner of it, De Morgan understood how modern pure mathematics in effect began with an unpleasant dispute over priority. Isaac Newton and Gottfried Leibniz had simultaneously discovered an important new method for handling infinitely small and infinitely large quantities—the calculus—leading to a long and involved debate between the two men and their respective supporters. However, in one of the few instances of Newton's inferiority, the Briton's system was more difficult to use than Leibniz's, and it concealed important insights from scientists who adopted it. Unfortunately for the progress of British science and mathematics, the British clung dogmatically to Newton's method. This mathematical parochialism distanced British researchers from significant Continental developments for more than a century. As Victorian mathematicians like De Morgan knew quite well, not until the Cambridge Analytical Society imported Leibniz's notation in the 1820s and 1830s did British science and mathematics reclaim the stature attained by the early members of the Royal Society.

How to ensure that history would not repeat itself? Critically, De Morgan's opposition to feuding within the ranks of the budding mathematical profession became inextricably related to his views about the nature of mathematics itself. As his dispute with Hamilton and his disgust with the early course of mathematical professionalization grew, De Morgan increasingly found relief in a new, more modest sense of his discipline. The age-old Platonic interpretation of mathematical knowledge as a bridge between the divine realm and the realm of matter would no longer be tenable. Mathematics did not consist of equations from God, De Morgan now argued—rather, it was a human construction, and thus less was at stake in mathematics than many people previously believed.

The way in which De Morgan's philosophy of mathematics evolved to serve his cultural and professional imperatives becomes clear when comparing his early vision of the discipline with his later descriptions. In a lecture on mathematics written in 1828 when he was only 21, De Morgan boldly portrayed the discipline as supremely exact and unlimited in its purview. The young mathematician extolled how mathematics, "this novel system of writing, this compendious language . . . has been from its peculiar structure, a never failing guide to new discoveries. . . . the study of this language, without reference to any of its applications, is instrumental in furnishing the mind with new ideas, and calling into exercise some of the powers which most peculiarly distinguish man from the brute creation."[132] With the fervor of a mathematical idealist, De Morgan proceeded to rank the "certain demonstrations of mathematics" as the highest form of knowledge, from which "the mind may safely descend . . . [to] the truths of metaphysics[,] of jurisprudence, or of political economy."[133] Furthermore, toward the end of the introductory lecture De Morgan advanced a somewhat authoritarian view of the discipline. The common perception that the student of mathematics "in every stage of his progress, is the judge of truth or falsehood" was quite erroneous, the young De Morgan claimed. Instead, he related the advice of "one of the most illustrious members of the fraternity of science, D'Alembert, a man equally distinguished as a mathematician, a metaphysician, and a natural philosopher." De Morgan continued, "His authority on this point is particularly valuable . . . 'go forward and faith will follow'. He recommends . . . [that a student] rely on the word of his instructor."[134] To the mature Augustus De Morgan, such an emphasis on respecting authority—and indeed much of his early characterization of mathematics—would become deeply problematic in a divisive social and professional context.

At nearly the same time De Morgan became interested in logic and spiritualism he began to reassess mathematical knowledge. In the 1840s he increasingly noticed that mathematicians grasp mathematical laws in a variety of equally valid ways. As De Morgan described his evolving sense of mathematics in a letter to William Whewell in 1844, "On the absolute substantive *reality* of all the primary truths of math[ematics] I have never had any doubt: but I have an idea that different people hold them by different hooks."[135] In *Formal Logic*, De Morgan publicly unveiled a more limited vision of mathematical knowledge based on this point. He explicitly agreed with Descartes that the knowledge of one's own thoughts was irrefutable, yet De Morgan believed that beyond this inner sphere we cannot achieve total certainty. Save *Cogito ergo sum*, he asserted, "We have lower grades of knowledge, which we usually call *degrees of belief*, but they are really *degrees of knowledge*."[136] In opposition to the mathematical idealists, De Morgan included among these lower forms the knowledge that $2 + 2 = 4$.[137] That is, rather than place the tenets of mathematics on a higher, divine plane (which could inspire absolute confidence in them), De Morgan lowered these principles to the realm of human knowledge. $2 + 2 = 4$ because *we human be-ings* say so, not because it is so in the mind of God.

Mathematics was a human construction, De Morgan therefore conceived by the late 1840s, and as such it was subject to human error. Of course, our confidence in $2 + 2 = 4$ is quite high. However, a small but very real possibility exists that it might not hold in every case. De Morgan proclaimed that even the great Isaac Newton's mathematical and physical laws, honored since his day as transcriptions from heaven, were less than truly certain. As he wrote about Newton's third law of motion, a touchstone of Enlightenment optimism about the power of human reason, "Nor do we know at this moment, as of necessity, that the proposition is correct. We have much reason to think that the law of equality of action and reaction is mathematically true: but, let it fail to the amount of only one grain in a thousand million of tons, and the proportion is not true, but only nearly true."[138] If we can question Newton, De Morgan implied, so much the worse for mere mortals. We may be *extremely* sure about a mathematical theorem, he concluded, but we can never be *infinitely* sure. $2 + 2 = 4$ for us, he argued, but what if there are sentient beings in some corner of the universe whose brains count logarithmically rather than arithmetically?[139] By 1851, De Morgan was telling William Whewell that he considered it unseemly the way "purely elementary truths may be lifted up to heaven" by mathematicians.[140]

At midcentury Augustus De Morgan thus came to think of mathematics as "a discipline of the mind"—the uncapitalized human mind, that is, not the divine Mind of mathematical idealism—and this attitude clearly guided his later research agenda.[141] For example, he regarded probability as a critical field of pure mathematics, because virtually all "truths" are merely probable (even though some of them may be exceedingly probable). Similarly, one can see why De Morgan became interested in the quantification of the predicate in logic. Words like "more," "less," and "somewhat" add crucial shades of gray to our descriptive palette. For De Morgan, very few Xs were *without a doubt* Y. Furthermore, if mathematics entailed human thoughts, not God's thoughts, then the intellectual history and conceptual foundations of the discipline required a great deal of attention. Studying the ideas of pioneering mathematicians was especially important, he believed, to clarify important conceptualizations that influenced subsequent practitioners.[142] A bibliophile and a biographer, De Morgan collected works by and about mathematical theorists from the ancient Greeks to the early Victorian age, and commented on a vast number of seminal treatises.

Moreover, in a strong parallel to his religious views, De Morgan thought that a precise comprehension of mathematical terms would help to rectify numerous misunderstandings. His study of the concept of infinity, for instance, exhibited this pragmatic concern with the terminology and psychology of mathematics. Unlike many contemporary philosophers and mathematicians who wrestled with the notion, De Morgan found infinity relatively untroubling following his downgrading of mathematical knowledge. As he noted almost casually in a series of letters to William Whewell in the early 1860s, mathematicians obviously could discuss and manipulate infinity even though they could not observe it in the real world nor create a mental image of it. "I have found out for some years," De Morgan told Whewell, "that I am a full believer in the infinitely great and small both, I mean in the *subjective reality* of both notions." The mathematician then described how he came to this position:

I cleared off much obscurity by a distinction which I find very faintly shadowed by the psychologists—that of a concept which has *image,* and a concept which has none. I can *image* a horse; I can't image the *right* to a horse.—but I can *conceive* it. I cannot image infinity but I can conceive it, that is, I recognize a notion with *predicates.* So that when a metaphysical writer says, as some have said, that we cannot conceive space to be finite, and are equally unable to conceive it as infinite, I say they ought to have said that we cannot conceive space as finite, nor image it as infinite. But neither can we *image* a million cubic miles— though we can conceive it, as proved by our knowing truth and false- hood about it.[143]

In other words, De Morgan conceded the impossibility of human beings grasping the notion of infinity (or even a very large quantity) through visualization. But mathematics did not require such a tremendous feat. Although "the infinitely small and the infinitely great are below and above our imagining power," De Morgan concluded in a final letter to Whewell, "they are concepts with attributes"—nothing more, nothing less.[144] Mathematics—even the advanced mathematics that dealt with elusive topics like imaginary numbers and infinite quantities—was merely an exploration of useful mental constructs.

By the 1860s, De Morgan had managed to develop a pragmatic strain of mathematical epistemology that he believed would halt "an epidemic in the air about mathematical thoughts" while vaccinating professional mathematicians against the debilitating effects of metaphysical and religious disputes.[145] The distance De Morgan had traveled from his 1828 lecture on mathematics is clear from his notes from the 1860s for a recast version of this introductory speech.[146] Whereas the earlier lecture made great claims about the potency and potential of mathematics, De Morgan now offered a well-calculated description of mathematics as being by nature nontranscendental and limited. He observed that mathematicians necessarily express their theorems in a language that cannot exist outside of the human mind: "Laws are mental enunciations . . . the names of diseases are not known to nature—nor are laws."[147] De Morgan contrasted this moderate view of his field with the extremism of other arbiters of truth in the Victorian age, highlighting the difference between mathematics and the "dogmatical theory of must and cant" in contemporary religion and philosophy.[148] "Modern philosophy," the mathematician disparagingly remarked, "returns (and more) to the old occult qualities . . . plenty of philosophy [is] not modest."[149] Mathematics, on the other hand, does not tolerate or permit such hubris.

Augustus De Morgan thus forged a modest philosophy of mathematics that he believed was inherently virtuous, and advantageous for the peaceful expansion of his discipline. Although ultimately opposed to philosophical idealism and the divine characterization of mathematics, the religious spirit so powerfully associated with pure mathematics in the early Victorian age nevertheless had its effect on him. By retaining mathematical idealism's claim to an ecumenical high ground, while at the same time discarding its grander religious connotations about mathematical symbols and concepts, De Morgan developed a new model for the last third of the nineteenth century. He shared many of George Boole's and Benjamin Peirce's spiritually derived imperatives, yet provided a secular approach to those issues that would become increasingly alluring to the growing ranks of professional mathematicians. In a charged cultural environment, De Morgan realized it would be best for everyone if religion and mathematics did not mix.

CHAPTER FIVE

Earthly Calculations
Mathematics and Professionalism
in the Late Nineteenth Century

For one it is the high, heavenly goddess;
For another it appears as a capable cow, which supplies him with butter.

—Friedrich Schiller

Augustus De Morgan's determination that mathematics should separate it-self from other realms of Victorian intellectual life, although shrewd, was not a self-fulfilling prophecy. To achieve this distinction required an aggressive promotion of the new isolationist ideal within mathematical circles as well as in the public sphere. As a boisterous critic and the first president of the London Mathematical Society, De Morgan lobbied brusquely, but effectively, toward this end during the last decade of his life. Other prominent mathematicians, aware of the opportunities and dangers associated with their professional standing within society, soon joined him. As their concerns increased in the 1860s and 1870s, a secular, research-oriented mathematics held the promise of a removed, "scientific," usefully incomprehensible sphere of experts.

In a sure sign of professionalization, mathematicians began to reexamine and redefine their discipline's foundations, categories, and relationships to other fields. Academic researchers, in particular, became more outspoken about the definition of mathematics and who should be called a mathematician. They forthrightly emphasized their training and certification, and formed more exclusive organizations and journals. Moreover, they worked diligently to distinguish their attitude and work from that of amateurs, and began to chide other mathematicians for "unprofessional" behavior, such as including theological rhetoric in technical papers. What may reasonably be called a "professional superego" developed—a set of behavioral rules internalized by mathematicians regarding what was proper and improper within their occupation.

Useful for this shift was a diminished sense of mathematics' intellectual and cultural ambition, anticipated in De Morgan's later thought. New advances in the understanding of mathematical symbols and concepts made the discipline seem less transcendental and less absolute in its proclamations. Scholars investigating number theory, algebra, geometry, and the calculus found weaknesses in the foundations of their fields. Once-secure mathematical structures now appeared to be more fragile than mathematicians had believed. Ancient texts such as Euclid's *Elements,* which remained central to mathematical education, as well as seminal modern works like Isaac Newton's *Principia,* seemed vulnerable for the first time.

Yet mathematicians were not obliged to abandon the traditional, religiously tinged philosophy of mathematics merely because new discoveries challenged the accepted wisdom. After all, they could have viewed advances such as non-Euclidean geometry in the same way some religious idealists viewed Darwin's theory of evolution—simply as further proof of the inexhaustible genius of the divine Mind. Furthermore, non-Euclidean geometry had been around for many years before mathematicians seized on it and broadcast its supposed impact on the ideology of mathematics. Why was a nontranscendental mathematical philosophy so attractive in the late Victorian era, and why were mathematicians so eager to remove religious innuendoes from their works? Why did this shift in the philosophy of mathematics occur in the second half of the nineteenth century, and not earlier or later? Although this transformation certainly emerged in part for internal reasons related to mathematical research and forms an important chapter in the progression of modern scientific thought,[1] the correlation between the ideological change and mathematical professionalization is conspicuous. A new generation of mathematicians who followed in the footsteps of Augustus De Morgan wrestled with issues surrounding their professional standing at the same time they struggled with the nature of their discipline.

Circle Squarers versus Professional Mathematicians

The fractious debate over the seemingly unworthy topic of circle squaring illustrates the difficulties that confronted professionalizing mathematicians in distinguishing their enterprise. Along with the duplication of a cube (finding a cube with double the volume of the original) and the trisection of an angle, producing a square with precisely the same area as a given circle was an ancient, and highly mystical, mathematical problem.[2] All three problems were bewitching because—unbeknownst to those who encountered these puzzles before the advances of modern mathematics—they involve the strange and elusive numbers called irrationals. Irrational numbers cannot be expressed as fractions involving whole numbers; put another way, they have endless decimal places with no pattern in sight. In the case of circle squaring, since the problem requires pinpointing the ratio between a circle's diameter and circumference, the irrational number the investigator bumps into is pi (π). Perhaps because of its extreme (in fact, total) difficulty—similar to the alchemist's hope of turning lead into gold—circle squaring offered its pursuers the dream of international fame in the discovery of an unknown quantity seemingly woven into the fabric of the universe. Great mathematicians from Archimedes to Descartes sought an answer to the problem, and intellectual luminaries such as Thomas Hobbes advanced their own solutions.

As mathematics progressed in the early modern era, however, it became increasingly clear that π was not like other numbers. Better calculations pushed Archimedes's remarkable estimate of between 3 1/7 and 3 10/71 far to the right of the decimal place, without a firm conclusion. In the late eighteenth century, researchers finally proved pi's irrationality. Johann Heinrich Lambert's work in the 1760s showed that $\pi/4$ could not be a rational number, and thus that π too was irrational.[3] Adrien Marie Legendre, the famous French mathematician and writer of textbooks, popularized this result in his influential primer *Éléments de Géométrie* (1794), which included a lucid description of the problem and declared the matter solved. The scientific academies had not waited for him—in the 1770s, the French Academy and the Royal Society of London refused to accept any more papers on circle squaring.

Despite Legendre's conclusion and the subsequent lack of interest in the subject among academics, a nagging doubt unfortunately remained. Irrational numbers come in two forms: algebraic and transcendental. The former can be expressed succinctly and totally (such as the square root of 3); the latter cannot. It was therefore conceivable that π might be a higher-order root or other more complex—though completely expressible—number. After all, the square root of 10 is tantalizingly close to the value for π. Why couldn't π turn out to be the cube root of some larger number or the sum of two roots? Despite confidence within the mathematical community, not until the relatively late date of 1882 did the German mathematician Ferdinand Lindemann prove that π was in the category of the transcendental, at last completely ruling out any circle-squaring solution.[4] Even so, Lindemann's highly intricate proof—using cutting-edge mathematical techniques—took considerable time to clarify and even more time to disseminate. The confusion surrounding the value of π consequently left the door slightly ajar to those who wished to carry on the ancient pursuit of equating a circle with a square.

Mathematics professors, however, had long given up on the endeavor. For them, the calculation of π had literally and figuratively become academic. Indeed, some early Victorian mathematicians raced to see who could calculate the greatest number of pi's decimal places. The calculations were painstaking and mind-numbing, with no obvious value other than bragging rights. Although even the most exact astronomical calculations used only the first eight or ten decimal places of π, mathematicians published each new mark with great fanfare. William Rutherford, a pioneer in mathematical education at the Royal Military Academy, Woolwich, broke the 200-decimal-place barrier in the early 1840s and doubled his total a decade later, only to be overtaken immediately by fellow Briton William Shanks.[5] In 1853, Shanks published a entire book on the subject entitled *Rectification of the Circle,* and by 1873 he had calculated π to over 700 decimal places.[6] In a way, he had made a career of it.

Just when circle squaring should have declined in the face of greater mathematical sophistication, it became something of a cottage industry among the growing ranks of Americans and Britons with a basic knowledge of mathematics. For example, in the early 1850s an American named John A. Parker pegged the exact value of π at $20,612/6,561$ and aggressively promoted his solution across the United States.[7] Although his proof was somewhat convoluted, Parker's appeal to common sense was attractive to a nonprofessional audience. Like many people, Parker failed to comprehend or accept both the form and intent of modern mathematics. He scoffed at the use of infinity and the infinitesimal, and derided the idea that a line could have absolutely no width—concepts which, after all, were distressing to many mathematicians and philosophers as well. Mocking the academic mathematician's use of infinity, Parker noted that "no explicit or satisfactory definition has ever been given" to the terms "infinite" and "infinity," and he thought it obvious that because these mathematicians used an infinite series to calculate the decimal places of π they were bound to generate a never-ending number.[8] "No mathematician ever did, or ever can understand" these terms relating to infinity, Parker wrote, and "if we reason from things which are *incomprehensible,* we reason of things which we know nothing about, and must fall into error."[9] Likewise, claiming from common sense "the truth of nature" that a line must have some width, Parker accounted for why his value for π was slightly more than the "accepted" value, as well as expressible in a fraction using whole numbers.[10] With his firm value for π and his even firmer stance against the academic pi-calculators, Parker ushered in a new vogue for circle squaring in the second half of the nineteenth century.[11] Victorian circle squarers tended to be upper-middle-class men with some engineering, medical, or technical expertise involving mathematics, but little knowledge of modern mathematical concepts.

The struggle of academic mathematicians for professional distinction in the 1860s and 1870s became clear in their feud with the new generation of circle squarers. This feisty group of amateur mathematicians troubled the professionals in several ways. First, they loudly represented a populist antagonism toward the academy at a time when universities were becoming central to mathematical professionalization. Furthermore, unlike academic mathematicians the circle squarers outlined a vision of mathematics that devalued extensive research and complexity in favor of revelations and simplicity. As Augustus De Morgan observed, circle squarers approached the problem "as a riddle: a thing to be thought out by some sudden stroke of mother-wit," an attitude that clashed with the growing (and professionally enhancing) perception of mathematics as a rigorous and time-consuming endeavor.[12] Finally, and perhaps most problematically, circle squarers maintained a strong bond between mathematics and religion. Most viewed π not merely as a numerical value, but as a pearl of wisdom from the mind of God. As academic mathematicians increasingly tried to extricate themselves from the realm of theology, this otherworldly conception of π became exceedingly troublesome.

The most prominent modern circle squarer was the Briton James Smith (1805–1872), who followed Parker with his own precise value for π. A well-known, wealthy Liverpool businessman, Smith sat on numerous boards, including the Liverpool Polytechnic, the Mersey Docks and Harbour Board, and the Liverpool Local Marine Board. He was also a member of the Liverpool Literary and Philosophical Club. Smith had studied geometry and dabbled in some other mathematical fields, such as algebra, which he occasionally used in his business. Moreover, he engaged in mechanical experiments relating to mining and energy production; in one major effort, he tried to make coal power more efficient.[13] From his expertise in applied mathematics, weights, and measures, Smith took a hands-on approach to the problem of circle squaring in the late 1850s. He cut pieces of cardboard and copper into circles and squares of different sizes, weighed them, and compared the sums. Through this process he developed a *feel* for the value of π, which he believed was exactly 3 1/16. However, a subsequent precise drawing of a circle and a square and an examination of the "mystical" properties of the numbers 7, 8, and 9 convinced him that he had been off by a sixteenth and that the true value of π was actually 3 1/8.[14] To this number he clung tenaciously until his death, pressing numerous mathematics professors to prove him wrong. Like the amateur biologist Robert Chambers, whose 1844 evolutionary treatise *Vestiges of Creation* achieved great notoriety and the derision of mainstream scientists, James Smith's work engaged the British public while haunting professional mathematicians.

Rather than ignoring him, several academic mathematicians waged a decade-long battle with Smith in the 1860s. At first professors such as Augustus De Morgan seemed more amused than offended. In his review of Smith's *The Problem of Squaring the Circle Solved* (1859), De Morgan concluded that the amateur's work was "a mode of meddling with unknown things which cannot do any harm, except to the spectator himself."[15] In private letters, De Morgan entreated Smith to renounce his circle-squaring solution. As Smith took their exchange public, however, De Morgan became increasingly worried. In Smith's publication of his conversations with the London mathematician, *The Quadrature of the Circle: Correspondence between an Eminent Mathematician and James Smith, Esq.* (1861), he disparaged De Morgan (who remains unidentified) for his academic entrenchment and aloofness.[16] Responding in *The Athenaeum* to the hostile *Correspondence*, De Morgan now adopted an entirely different tone: "[Smith] deserves the severest castigation: and he will get it."[17] In his long review of the book and subsequent publications, De Morgan began to escalate his attacks on Smith and his ilk.

The UCL mathematician clearly had become concerned about the effects of Smith's populism on the general public's view of his discipline. As De Morgan worried aloud on the pages of *The Athenaeum*, "To the mathematician we have nothing to say: the question is, what kind of assurance can be given to the world at large that the wicked mathematicians are not acting in concert to keep down" Smith and other circle squarers.[18] Seeking public support, Smith in turn sent letters to various newspapers, including the Liverpool *Albion,* decrying De Morgan's increasingly vicious responses. De Morgan then published a simple proof of Smith's error (provided by William Rowan Hamilton) in *The Athenaeum,* to which he attached another sharp barb: "We give [the proof] in brief as an exercise for our juvenile readers."[19] In yet another populist salvo, *A Nut to Crack for the Readers of Prof. De Morgan's 'Budget of Paradoxes'* (1863),[20] Smith updated his original proof and uncharitably referred to De Morgan as an "elephant of mathematics." Smith filled his books with as much antiacademic bile as mathematics. His technical passages often devolved into critiques of the growing academic monopoly on mathematical truth.[21]

Following James Smith's affront to the "Eminent Mathematician," mathematicians quickly closed ranks around Augustus De Morgan. Smith encountered this professional solidarity when he submitted his circle-squaring result to the mathematics section of the British Association for the Advancement of Science at its Manchester meeting in 1861. He had slyly proposed his value for π in an official-looking paper entitled "On the relations of a square inscribed in a circle," but a review committee uncovered its true intention and disallowed its presentation. Later, the mathematician and astronomer George Biddell Airy summarized the growing professional antagonism toward circle squarers like Smith while delineating the proper territory and demeanor of mathematics. In his presidential address to the section, Airy proclaimed that mathematicians should subsequently "consider themselves as treating questions strictly of science." The brevity of professional conferences mandated the exclusion of material deemed extraneous or incomprehensible "by the majority of the persons present." Every real mathematician, Airy argued, regarded circle squaring as "a mere loss of time . . . [that] has been rejected by the learned in all ages," and it was essential that "subjects of that kind should not be admitted." Aside from their defective mathematics, Airy also found the contentiousness of Smith and other circle squarers offensive. He demanded that "personality of every kind . . . be strictly eschewed" at conferences in favor of an even-keeled, respectful, professional comportment.[22] In a testy and uncompromising way, Airy thus drew a sharp distinction between these interlopers and true mathematicians.[23]

At the same moment, P. H. Van Der Weyde, a professor at Cooper Union in New York City, confronted the American circle squarer John Parker. In his pamphlet *The Quadrature of the Circle Demonstrated to be Perfectly Solu-*

ble by Modern Mathematics (1861), Van Der Weyde scoffed at the treatises of "that unfortunate class of persons" who still attempted to square the circle, concluding (somewhat contradictorily) that "it would be wasting time and paper to say anything about such a book; it would do nobody any good, and surely not its author."[24] Van Der Weyde's elitism undoubtedly irked Parker. The professor found it amusing—and significant—that circle squarers often had such commonplace names, patronizingly calling the bunch "Mr. Smith, Jones, Parker, et al."[25] In other words, circle squarers were average men improperly dabbling in a complex and recondite science that was the province of far greater intellects.

Like James Smith, Parker responded by accusing the academics of fostering an unsavory dogmatism. Professors unthinkingly worship idols as much as everyone else, he emphasized. "I confess to an abundant surprise at finding, that the professors of our own day and in our own country particularly, have received what Legendre and a few others have said, as *established facts,* and have adopted their opinions without investigation," Parker declared after a decade of goading Van Der Weyde and other academics.[26] In a way, Parker was right. Although the proof of the irrationality of π was not a social construct, few mathematicians checked the validity of each of its steps. Mathematics professors simply saw the matter as solved, and they had already moved on to newer problems. Recapitulating the entire history of the field to do research on its frontiers made no sense.

Parker read this pragmatic attitude as haughtiness, however, an aloof complacency at odds with the demands of their discipline. From his perspective, academic mathematicians converged on a single viewpoint and worked together to reject alternative solutions, when they should be striving on many different paths toward great new truths.[27] "If we examine into the history of the progress of science we shall find, that the great stronghold of this mental disorder has always been found within the walls of the academy," Parker wrote of this homogeneity among professors.[28] Formal education apparently discouraged independent thinking and even small variations in notation.[29] "From this cause I have found the Professors as a body, though learned in the received theories, to be among the *least competent* to decide on any newly discovered principle. Their interest, education, pride, prejudice, self-love and vanity, all rise in resistance to anything which conflicts with their tenets, or which outruns the limits of their own reasoning," Parker concluded, implying that freethinking amateurs such as himself thus had a crucial role to play in the vanguard of mathematical re-search.[30]

Although John Parker's writings drifted toward paranoia, he undoubt-edly encountered academics' increasingly serious intention to draw a line between amateur and professional mathematics in the 1860s and 1870s. In *A Budget of Paradoxes*, Augustus De Morgan expressed the sentiments of his colleagues by projecting the mistakes of circle squarers onto an entire class of amateur mathematicians. With characteristically little tact, he sharply di-vided legitimate mathematicians from the James Smiths of the world, whom he referred to as "pseudomaths." "The *pseudomath* is a person who handles mathematics as the monkey handled the razor," De Morgan snidely remarked, "The creature tried to shave himself as he had seen his master do; but, not having any notion of the angle at which the razor was to be held, he cut his own throat. He never tried a second time, poor animal! but the pseudomath keeps on at his work, proclaims himself clean-shaved, and all the rest of the world hairy."[31] Instead of technically critiquing Smith's work, De Morgan now patronizingly rebutted the amateur by casting him as a comically incapable primate. He also echoed Airy's attack on the personal-ity of circle squarers by portraying Smith as a stubborn, slightly crazed cler-gyman—"the Supreme Pontiff of cyclometers, the vicegerent of St. Vitus"—and circle squarers in general as beyond the pale, as "the wild Welchmen of geometry" who operated without "the rules of good society."[32] The contrast between the well-behaved professional mathematicians and the unstable, unwelcome amateurs could not have been clearer.

Such name-calling by the president of the London Mathematical Society and other mathematicians, if occasionally juvenile, seems to have had an ef-fect. By the end of the 1870s, the number of circle-squaring treatises from "respectable," technically skilled individuals had dropped off noticeably. Tellingly, one of the few published proofs from the late 1870s, *The Secret of the Circle* by Alick Carrick (1876), is almost certainly pseudonymous—prob-ably the work of F. B. Playfair, a fellow of the Royal College of Surgeons, who wrote the preface.[33] Of course, attempts to square the circle continued, but the problem's new pursuers had far less mathematical knowledge and prominence than Parker, Smith, or Playfair. A typical post-1870s tract was Edward Moore's *Geometrical Science* (1890), which exclaimed in capital let-tering on the cover that the "great problem which has baffled the greatest philosophers and the brightest minds of ancient and modern times, from

Phythagoras [*sic*], five hundred years before Christ, to the present day, has now been solved by an humble American citizen, of the city of Brooklyn."[34]

Wishing never again to repeat the uncomfortable experience of the 1861 British Association for the Advancement of Science meeting, new mathematical organizations erected greater barriers to circle squarers and amateurs in general. In particular, these institutions raised the importance of an academic affiliation. The London Mathematical Society's original twenty-seven members were almost entirely faculty or graduates of University College, London. The 1884 LMS membership roll, although having greatly expanded to a total of 170 mathematicians (excluding foreign members), contained only twenty-one scholars who were not affiliated with a university, and twenty-one members who did not hold a university degree (seven of whom worked at a university despite this handicap). Subtracting government and military position holders, independent mathematicians comprised less than a tenth of the 1884 LMS membership.[35] The Edinburgh Mathematical Society, founded in 1883, followed the model of its southern cousin. Established "primarily in connection with the University [of Edinburgh] . . . for the mutual improvement of its members in the Mathematical Sciences," it restricted its membership to those who were taking or had taken advanced mathematical classes at the University, who were honors graduates from another British university, or who were "recognized" mathematics teachers—that is, recognized by the members of the organization.[36] Similarly, of the 225 members of the nascent New York Mathematical Society in 1892, only fifty-one labored outside of a university setting. Of these fifty-one mathematicians, twenty-two were in business (mostly insurance actuaries), eleven worked for the government, and five held positions in secondary education. Thus only thirteen members were independent scholars, and most were so well known as to be beyond reproach (e.g., the eminent philosopher Charles Sanders Peirce, son of Benjamin Peirce).[37] By the end of the century, the umbrella American Mathematical Society consisted almost entirely of professors and doctorate holders.[38] The academy therefore became increasingly central to the mathematical profession in the last third of the nineteenth century—one could be a historian or even a chemist somewhere else, but to be a recognized "mathematician" virtually always meant that one held a position at a university or college.

Puzzles versus Research

Excluding circle squarers from professional organizations was far easier for mathematicians than overcoming their popular sense of mathematics as a puzzle-solving endeavor. For generations, after all, the Mathematical Tripos at Cambridge University—the standard-bearer of mathematics education in Great Britain—had reinforced this common perception. The exam, fundamentally unchanged since the middle of the eighteenth century, and until 1850 a requirement for the B.A. degree, essentially consisted of problems derived from the first four books of Euclid's *Elements* and parts of Newton's *Principia*. Often written in short sentences rather than diagrams or equations, typical questions asked, "Given the three angles of a plane triangle, and the radius of its inscribed circle . . . determine its sides" or "How far must a body fall internally to acquire the velocity in a circle, the force varying $1/D^2$?"[39] These questions varied little from year to year, testing less the ability to think creatively than the talent for manipulating conventional, existing techniques.

Furthermore, many early mathematics periodicals characterized the discipline as a source of diverting puzzles. Most prominent among these journals in Great Britain were the various permutations of *The Gentleman's Diary* and its sibling *The Lady's Diary*, both dating from the eighteenth century. A combination of astronomical knowledge (phases of the moon, etc.), poetic riddles, correspondence, and mathematical problems, the *Diaries* promised their elite readers the same well-rounded stature the educated classes associated with the Cambridge Tripos. The first volume of *The Gentleman's Diary* advertised itself as "containing many useful and entertaining particulars, peculiarly adapted to the ingenious Gentlemen engaged in the delightful study and practice of the Mathematics."[40] Likewise, *The Lady's Diary* highlighted that it "contained new improvements in Arts and Sciences, and many entertaining Particulars: designed for the use and diversion of the fair sex."[41] Both journals published several fairly challenging mathematical problems in each issue, requiring some knowledge of geometry and algebra but not modern developments such as the calculus. American journals from the early nineteenth century, such as *The Mathematical Diary* and *The Mathematical Companion*, had a similar orientation.[42]

Midcentury shifts in academia and mathematical print culture, stressing the depth and seriousness of the discipline, opposed this portrayal of mathematics as entertaining puzzles. Pressure to modernize the mathematics curriculum, which had begun in the 1810s with the Cambridge Analytical Society (founded by George Peacock, John Herschel, and Charles Babbage), and which was adopted by William Whewell as a *cause célèbre* in the 1840s, finally came to fruition. A board set up to reevaluate the Tripos split the exam into simpler and more advanced topics and added a host of subjects to supplement (or crowd out, according to detractors) the essential texts of Euclid and Newton. In addition, examination questions increasingly consisted of mathematical notation rather than words. This reformation essentially made advanced mathematics an independent track of study rather than a requirement at Cambridge University. To gain honors in the new Moral Sciences Tripos or Natural Sciences Tripos, a Cambridge student only needed to take the first, easier part of the mathematics examination.[43] A subsequent revamping of the mathematics curriculum to include the most recent and complex topics quickly diminished the number of students concentrating in the field.

These momentous changes in mathematical education swiftly moved from Cambridge to other universities. Mathematics thus professionalized during a period of sharp *decline* in the population of students in the discipline; in the 1870s, the number of honors graduates in mathematics actually reached a 40-year *low* in British universities. This trend was apparent not only at Cambridge, but also at Oxford and the University of London.[44]

At the same time, a new breed of mathematics journals arose that helped to turn the discipline sharply away from puzzle solving and other characteristics associated with amateurism. Following the lead of the German *Journal für die reine und angewandte Mathematik* (Berlin, founded in 1826) and the French *Journal de Mathématiques Pures et Appliquées* (Paris, founded in 1836), the members of the Cambridge Analytical Society and other elite British mathematicians established the *Cambridge Mathematical Journal* in 1837. Like its Continental siblings, the *Cambridge Mathematical Journal* focused on advanced research, and with its minuscule circulation and unsigned articles it was largely an internal dialogue among an extremely small set of mathematicians.

In the mid-1850s, however, the *Cambridge Mathematical Journal* gave way to the broader and more ambitious *Quarterly Journal of Pure and Applied Mathematics.* The *Quarterly Journal,* edited by James Joseph Sylvester and N. M. Ferrers with the assistance of George Gabriel Stokes and Arthur Cayley, had a grandiose sense of itself and the budding mathematical profession. In their preface to the first volume, the editors exalted in the "new powers" of mathematics and announced that Britons could no longer "remain in-debted to the courtesy of the editors of foreign Journals . . . [for] the rapid cir-culation and interchange of ideas by which the present era is characterised."[45] By appealing to "the hearty co-operation of . . . fellow labourers in mathe-matical pursuits," they envisioned a professional community that would move forward as a unified force.[46] This serious, progressive tone prevailed throughout the journal. Puzzles and rhymes were out; theoretical research and technical language was in. Moreover, by focusing on "the requirements of that important class of readers, from whom the ranks of Mathematicians must hereafter be recruited," the editors of the *Quarterly Journal* established a strong link between university training and a career in mathematics.[47]

Responding to their mentors' call, elite university mathematics students founded the first graduate student journal, *The Oxford, Cambridge and Dublin Messenger of Mathematics,* in 1861. The *Messenger's* table of contents proudly displayed the contributors' credentials, with bachelor's and master's degrees prominently affixed to each student's name. In addition, like their elders the students distanced themselves from puzzles and basic problem solving, in-stead emphasizing that only novel research belonged in a true mathematical journal. "The solving of problems, unless they are of a high order of origi-nality and conception, can scarcely be looked upon as original investigation at all," the editors of the *Messenger* proclaimed in the introduction to the first volume.[48] Toward this end, the editors explicitly cautioned their audience that they would "exercise a somewhat strict censorship" over submissions in-volving "riddle-solving" and "puzzle-working."[49]

Other new journals geared toward mathematics students contained similar caveats about these "amateur" pursuits. *The Mathematical Monthly,* printed in the United States and Great Britain beginning in 1859, contained a few puzzles, but its editor also conspicuously demeaned such diversions in the introduction to the first issue: "Problems of the highest grade, especially if they are likely to lead the investigator into a comparatively new field, or de-velop methods or important practical results, may occasionally be published as *challenges;* but generally, we think it advisable to publish the solution of such problems at once."[50] Mathematical puzzles were not for amusement, in other words, but for hard-nosed pedagogy. Mathematics education might be different from advanced research, but it should maintain the same serious tone.

The *American Journal of Mathematics,* which began publication in 1878 under the auspices of the new Johns Hopkins University (a school instrumental in the rise of the doctoral degree as a professional stamp), strictly disallowed mathematical puzzles or basic problems. Indeed, the associate editor, Hopkins' William E. Story, did not mince words in his preface to the periodical's first volume, announcing, "It is to be understood that there will be *no problem department* in the Journal." He directed problem solvers to *The Analyst* and *The Mathematical Visitor,* two amateur magazines founded in the wake of the new professional journals.[51] To buttress its exclusivity, the *American Journal of Mathematics* was considerably more expensive than a periodical intended for a popular audience: five dollars a volume, a tremendous amount for the day (and quite prescient of science journal pricing in the twenty-first century).

The successful division of mathematical print culture into professional and nonprofessional strata is especially clear in *fin de siècle* amateur magazines such as the *Mathematical Gazette* and *The American Mathematical Monthly,* both founded in 1894. Two members of the New York Mathematical Society, B. F. Finkel and J. M. Colaw, edited the latter, which they specifically geared toward the audience left behind as new mathematics periodicals catered to an increasingly rarefied circle. "At the present time there is no Mathematical Journal published in the United States sufficiently elementary to appeal to any but a very limited constituency," Finkel and Colaw noted in the first issue.[52] Perhaps overestimating the appeal of the *Monthly,* the editors hoped to make the magazine "the most interesting and popular journal published in America," and toward that end they included puzzles and simple problems.[53] Yet Finkel and Colaw also felt the need to put such diversions down as "one of the lowest forms of Mathematical research . . . [which] in general . . . has no scientific value."[54] With their professional status secure, the editors thought they could safely reintroduce playful mathematical forms—though with an explicit warning that the public should not confuse riddles with higher mathematics. For years, professional mathematicians like Finkel and Colaw had drawn a line between puzzle solving and the true pursuit of their discipline; now this distinction existed on library shelves as well.

The Continuing Relationship Between Mathematics and Religion

Lurking behind the antiprofessional diatribes and puzzle-solving disposition of amateur mathematicians like the circle squarers was another powerful sentiment: a yearning for a simplicity and magic in numbers that academic mathematicians increasingly seemed eager to dispel. Indeed, the method and result of the brute computation of π were strongly antithetical to these ideals. At 500 or 1,000 nonrepeating decimal places calculated by hand, π begins to look a little drab and unworthy of reverence. It certainly lacks the aura of the Bible's value of 3 (I Kings 7:23, II Chronicles 4:2)[55] or the ancient Egyptian value of $(4/3)^4$,[56] not to mention the various fractions advanced in the nineteenth century. For Victorian circle squarers who believed mathematics should be about mystical insight and perfection, not ignoble approximation, this was a grossly unsatisfactory result. Their dissatisfaction was substantially the same as those who found Charles Darwin's adequate, but imperfect, adaptation of species through natural selection to be utterly inferior to the perfect structures described by natural theologians such as William Paley. As many Victorians understood (including Darwin himself), without such perfection envisioning an omnipotent and beneficent Author of Creation became difficult.

James Smith's *The Problem of Squaring the Circle Solved*, which enjoyed the greatest recognition of any circle-squaring effort, may have tapped into a widespread desire for an ideal vision of mathematics.[57] Summarizing his aesthetic of mathematics, Smith argued forcefully, "Perfection is superior to the nearest approximation thereto, and certainty is at all times better than doubt; and if by [my] discovery we obtain the one and get rid of the other, which with reference to this subject has not hitherto been attained, surely something must be gained. It at least illustrates what is universally found to be true in nature, that in all her operations she is simple, consistent, harmonious, beautiful, and perfect."[58] Along these lines, Smith advocated other precise constants related to his value for π. He asserted that the earth was a perfect sphere measuring exactly 8,000 miles in diameter and 25,000 miles in circumference, rather than the slightly flattened shape and less rounded measurements held by modern science; that the other planets were similarly perfect and revolved around the sun in circular rather than elliptical orbits; and that the year was precisely 365 1/4 days rather than 365 days, 5 hours, 49 minutes, and 12 seconds.[59] Furthermore, Smith appealed to scientists who used mathematics extensively, such as astronomers, to let their work remain a lofty pursuit of perfection rather than become a vocation characterized by the drudgery of arduous (and only approximate) computations. If an astronomer accepted the value of π as 3 1/8, he "will be relieved from a large amount of mental labour, in the long and tedious calculations

he has continually to engage in. And by means of it the delightful study in
· which he is occupied will become more and more a labour of love."[60] Smith
worried that mathematicians were purging their noble discipline of its divine purpose and transcendental flawlessness.

The American circle squarer John Parker also challenged this apparent
degradation of mathematics by promoting a greater metaphysical view of
the discipline. "What are numbers?" he asked in the epilogue to his *Quadrature of the Circle*. "Before creation began," he answered, "numbers had no
existence, except in the infinite eternal One."[61] Parker wondered what was
happening to the traditional view of "every mathematician, geometer, or astronomer, of eminence, to whom the world owes all that is known on these
subjects, that the quadrature of the circle *was*, and *is*, an elementary truth,
necessary to be known for the perfection of mathematical and astronomical
science."[62] The academic value of π was perplexing and disillusioning. What
was happening to "the most perfect of *all* the sciences" that academic mathematicians should accept such an unsatisfactory approximation?[63]

Indeed, most circle squarers saw themselves—perhaps justifiably—as
merely reflecting the religious idealism of many early Victorian mathematicians. For example, William Alexander Myers, the president of the
eponymous Myers Commercial College in Louisville, Kentucky, espoused
an idealist philosophy of mathematics in his ambitious book *The Quadrature of the Circle, the Square Root of Two, and the Right-Angled Triangle* (1873).
"By the aid of the power which we obtain from the science of numbers," Myers wrote in the introduction, "we can reason mathematically and truly far
beyond what we can see."[64] In other words, Myers thought he was simply
following in the footsteps of the mathematicians who had prognosticated
the size and position of Neptune. For him, as for other Victorian circle
squarers, mathematics (and, in particular, its greatest accomplishments)
came into contact with the mind of the grand Designer. Myers declared that
"it may be said with truth that the circle, the square, and even the triangle
are emanations from the divine intelligence, as well as the science of numbers, by the aid of which they are measured."[65] For this reason, he found inspiration from Scripture as well as from mathematical tracts. Myers discussed his use of the Douay Bible along with "De Morgan on the Law of
Probabilities" in the composition of his proof, and mentioned how his result matched number patterns in the Old Testament. Only one conclusion
could be reached: "[God] must have used such numbers as in His infinite
wisdom would best accomplish that result, and these numbers *must* be such

as is commonly found in all His works."[66] In *The Square of the Circle* (1893), Rufus Fuller likewise claimed a biblical inspiration for his work: "On learning the inner meaning of Scripture numbers and other arcana never made known before . . . the inception was formed of entering upon this present work, pertaining to measurements and problems in the natural world."[67] In addition to its other connotations, the squaring of the circle thus had a religious overtone—it sustained the belief that mathematics could access wondrous, divine truths.

At the same time circle squarers were using religious arguments to justify their mathematics, clergymen were using mathematical arguments to justify their religion. These clerics, like the circle squarers, reasonably could appeal to precedent: prominent early nineteenth-century academic mathematicians who made theological points with their discipline. Possibly the best-known academic mathematician who spoke his mind on theological issues was Charles Babbage (1791–1871), holder of the prestigious Lucasian chair at Cambridge University at the beginning of the Victorian era. Babbage's *Ninth Bridgewater Treatise* (1837) brazenly employed mathematical principles to support religious doctrine. For instance, he used equations from Laplace's *Théorie Analytique des Probabilités* and Poisson's *Recherches sur la Probabilité des Jugements* to counter David Hume's repudiation of miracles, calculating the probability that so many witnesses could have been mistaken about seeing Jesus rise from the dead. Needless to say, the mathematical computation showed an incredibly slim chance (a trillion to one) that the resurrection was a mass illusion, as Hume had implied.[68] To add a seal of approval to his proof, Babbage underlined that the probability theories he used "are established, and they are not merely undisputed, but are admitted by other writers of the highest authority on this subject."[69] Babbage also buttressed William Paley's argument from design (for the existence of God) using inchoate versions of the mathematical series he would later incorporate into his calculating engine.[70]

If the Lucasian Professor could make such claims, other early nineteenth-century academic mathematicians unsurprisingly felt comfortable mixing theology with mathematics in their writings. Olinthus Gregory (1774–1841), who for three decades held the mathematics chair at the Royal Military Academy, Woolwich, clearly saw no conflict between his religious and mathematical proclamations, and his publications flowed seamlessly back and forth between the two subjects. Noting that "in several parts of pure and mixed mathematics, there are numerous incontrovertible propositions, which are, notwithstanding, incomprehensible," Gregory found the lessons of pure mathematics extremely useful when discussing the existence, yet inscrutability, of God.[71] In his *Letters on the Evidences, Doctrines, and Duties, of the Christian Religion* (1811), Gregory explored some of these "mysterious" elements of mathematics in detail, including irrational numbers, imaginary quantities, and infinity, "in order to recommend that similar principles . . . be adopted, when *religious* topics are under investigation."[72] "We cannot comprehend the nature of an infinite series," he emphasized, "[and] in like manner . . . we cannot, with our limited faculties, comprehend the infinite perfections of the Supreme Being, or reconcile his different attributes, so as to see distinctly how 'mercy and peace are met together, righteousness and truth have embraced each other;' or how the Majestic Governor of the universe can be every where present, yet not exclude other beings."[73] In short, the grand notions of higher mathematics provide compelling parallels to divine knowledge. Pure mathematics, like faith, revels in the Unknowable.

Similarly, in 1839, Oliver Byrne, a professor of mathematics at the College of Civil Engineers, London, published *The Creed of Saint Athanasius Proved by a Mathematical Parallel,* which lamented the decline of orthodox faith and sought to remedy this deterioration through mathematics. "It is appalling to see some hundreds of thousands depart from the church of Christ and become Deists or Atheists, or remain dissatisfied, or deterred to examine; because the powers of their minds are not comprehensive enough to understand how that most beautiful combination of heavenly power and grace can exist, as is set forth in the Athanasian Creed," Byrne wrote at the beginning of his treatise.[74] Following this philippic, Byrne elucidated the complex orthodox description of the Trinity—the simultaneous unity and distinction of the three manifestations of God—through pure mathematics. First he equated each element of the Trinity (the Father, the Son, and the Holy Ghost) to the mathematical sign infinity, since the Athanasian Creed considered all of them to be unlimited. Byrne then erected two vertical columns: the left containing the English Book of Common Prayer translation of the *Quicunque Vult* (the traditional description of the Athanasian Creed), the right containing parallel mathematical equations involving infinity that purported to establish the truth of the statements on the left. Since the Father, the Son, and the Holy Ghost were all equal to infinity, and since infinity added to or multiplied by itself is infinity, Byrne's proof of the Athanasian Creed was not especially complicated. On the left the theological phrases mounted: "So likewise the Father is Almighty, the son Almighty:

and the Holy Ghost Almighty. And yet they are not three Almighties: but one Almighty. So the Father is God, the Son is God: and the Holy Ghost is God. And yet they are not three Gods: but one God. So Likewise the Father is Lord, the Son is Lord: and the Holy Ghost is Lord. And yet not three Lords: but one Lord . . ."[75] On the right the various infinities representing the Father, the Son, and the Holy Ghost ((∞)^f, (m∞)^s, and (∞)^g, respectively) went through their paces: "(∞)^f + (∞)^g + (m∞)^s = ∞ = (∞)^f = (∞)^g = (m∞)^s, and ∞ = (∞)^f = (∞)^g = (m∞)^s = (∞)^f + (∞)^g + (m∞)^s . . ."[76] To some readers, *The Creed of Saint Athanasius Proved by a Mathematical Parallel* may seem like an absurd tautological comedy, yet it was composed with utmost seriousness by a devout professor of mathematics.

Clergymen were quick to follow the example set by Babbage, Gregory, and Byrne. "When a very young man, I was frequently exhorted to one or another view of religion by pastors and others who thought that a mathematical argument would be irresistible," Augustus De Morgan recalled disapprovingly in his old age.[77] *The Two Estates; or Both Worlds Mathematically Considered* (1855), a treatise circulated in London comparing earthly existence with the heavenly afterlife, typified this trend. The treatise's anonymous writer argued that since x, the value of eternal bliss, was infinitely greater than a, the value of all possible happiness in this life, then $x + a = x$, meaning that a is irrelevant. The author also noted with satisfaction how his method of "neglecting infinitely small quantities" was one and the same as the method "Sir Isaac Newton was indebted to for his greatest discoveries."[78] Speaking in front of an audience at the Boston Lyceum, the American divine Henry Ware Jr. also found inspiration in the infinite sense of mathematics: "There is the doctrine of the *asymptotes*, which teaches that a right line may approach forever to a certain curve, and yet, though infinitely extended, will not touch it . . . what an idea for the imagination to dwell upon is this!"[79] In its vertiginous balance between awe-inspiring inscrutability and human comprehension, pure mathematics provided a modern-day means for theologians like Ware and the author of *The Two Estates* to uphold faith.

The use of mathematical concepts in sermons and theological treatises perhaps reached its apogee in the work of the Church of Ireland clergyman Tresham Gregg (1799–1881). In the 1850s, this Dubliner became fascinated with pure mathematics and mathematical logic, and he sought to apply them to religious concerns. The result was his remarkable 1859 monograph *Novum Organum Moralium,* helpfully subtitled *Thoughts on the Nature of the*

Differential Calculus, and on the Application of its Principles to Metaphysics, with a View to the Attainment of Demonstration and Certainty in Moral, Political, and Ecclesiastical Affairs.[80] Gregg reduced human beings and the Church to sets of algebraic variables in a manner that recalled Boole's *The Laws of Thought* and De Morgan's *Formal Logic.* Each human being became a product *cex*, where c = man's calling; e = man's habitudinal attitudes, and x = man's special individuality; likewise, Gregg expressed the Church as $A\delta X$, where A = apostolic origin; δ = doctrines; and X = the congregation of the faithful.[81] He even established variables for the traits of God: "He is the giver of all knowledge (k) and the source of all virtue, (h); the creator of intellect (i) and the bestower of power physical (p)."[82] Gregg then used mathematical laws, including axioms from algebra and the calculus, to find the answers to such vexing doctrinal questions as the value of the clergy, the correct structure of the Christian family, and the accurate interpretation of the Bible. One equation, for instance, encapsulated the growth of David from shepherd to king:

$$\frac{du}{de} = ce\frac{dx}{de} + ex\frac{dc}{de} + cx$$

Gregg believed this formula showed irrefutably "that the increase of David's educational excellence or qualities—his piety, his prayerfulness, his humility, obedience, &c., was so great, that when multiplied by his original talent and position, it produced a product so great as to be equal in its amount to royalty, honour, wealth, and power."[83] *Novum Organum Moralium* ends— much like a student's workbook—with some study questions for the reader to solve using Gregg's religio-mathematical method. For example, "What moral truth does the truth, that a number multiplied by its inverse is always equal to unity, illustrate?" and "Show how the analogies of pure science coincide with the ideas of the eternal retribution of good and evil, and are in contravention of the idea that man should sink into a state of annihilation."[84] From Tresham Gregg's perspective, mathematics was a special language with unparalleled potency. How could the clergy ignore it?

Engaged in a mutually beneficial relationship, religion and mathematics thus remained strongly affiliated in the first half of the Victorian era. Many believing mathematicians saw no reason to shy away from theological discourse. Indeed, pure mathematics gave them useful material for religious suasion. Similarly, clergymen made forays into mathematics to borrow analytical methods whose rigor lent the appearance of unassailability to their declarations, and concepts whose transcendental aura made them highly attractive for discussions of God. Together, these mathematicians and theologians continued the tradition of their ancient, medieval, and early modern forerunners, sustaining an understanding of mathematics inseparable from the sphere of faith.

The Secularization of Mathematics

This close association between mathematics and religion became more troubling to academic mathematicians as the nineteenth century wore on, however, as British society experienced perennial (and frequently predictable) sectarian disputes. The Established Church and nonconformists had long been at odds; feuds now erupted within denominations as well. Doctrinal disputes rent the Church of England, the Unitarians, and even the Quakers. Everywhere a hardening of positions occurred that made ecumenical dialogue difficult within and between sects.[85] And in the same year Tresham Gregg published his *Novum Organum Moralium,* Charles Darwin released his *Origin of Species,* producing even more public dissension on spiritual matters. The idealism of George Boole, who dreamed of bridging religious divides, appeared increasingly naïve.

Vanguard mathematicians continued to witness firsthand the destructiveness of factionalism. In 1866, for instance, a position opened for a professor of "mental philosophy and logic" (i.e., psychology) at University College, London. One of the candidates was James Martineau, a well-known and well-qualified theorist who also happened to be a liberal Unitarian. His candidacy quickly advanced, broadly supported by the faculty senate, but he suddenly encountered opposition when members of the college council learned of his religious views. On the one hand, aristocratic Anglicans on the council balked at his denomination and set themselves against his appointment. Meanwhile, another group of professors found Martineau's theories *too* religious, because he held that all psychology had its basis in a sole moral Governor. These "philosophical men," as Augustus De Morgan called them (i.e., phenomenalists and Positivists), disagreed with such a notion, preferring less of a theist in a field they considered nontheological in nature. They found a crucial ally in the powerful vice chancellor (and later president) of UCL, George Grote. Grote, best known for his *History of Greece,* had been strongly opposed to a clerical professor of mental philosophy from UCL's inception. Thirty years earlier he had resigned his chair on the original college council after a clergyman received such an appointment. At the August 1866 meeting of the council, Grote put forward a motion to quash the candidacy of any religious minister on the principle that such candidates contradicted the religious neutrality of the college.[86] By the end of the year, the aggressive Grote, the secularists and the Church of England antagonists forged an unusual but effective alliance that sunk Martineau's candidacy.[87]

The council's decision incensed many Unitarian and ecumenical sympathizers. It seemed clear to De Morgan, among others, that the dark specter of sectarianism had returned to UCL. "Univ. College has abandoned its principle," he fumed to a confidant in the fall of 1866, "and has pronounced the religious notions and action of a candidate a disqualification for a chair."[88] The peculiar coalition of the orthodox and "philosophical" factions sickened De Morgan. "Mr. M. was rejected because he was too far from orthodoxy to please the priests, and too far from atheism to please the philosophers . . . he was offered up to the Janus Bifrons of expediency, each member of the majority of the Council choosing the head of the idol to which his offering was to be made," De Morgan complained to his friend William Heald.[89] His profound disgust over the Martineau affair eventually spurred De Morgan to resign from UCL for a second and final time. Many former students implored him to stay; when he told them he could not, they asked him to sit for a portrait to be mounted prominently in the college. De Morgan's emotional response to their offer displayed his tremendous anger about the situation:

I am asked to sit for a bust or picture, to be placed in what is described as "our old College." This location is impossible; our old College no longer exists; it was annihilated in November last. The old College, to which I was so many years attached by office, by principle, and by liking, had its being, lived, and moved, in the refusal of *all religious disqualifications*. Life and soul are now extinct. . . . I now wish my life had been passed in any other institution. I have worked under the conviction that I was advancing a noble principle, until every letter in the sentence, "Augustus De Morgan, Professor of Mathematics in University College, London," stands for 234 hours of actual lecturing, independent of all study and preparation. And all this under a banner which is now shown to have been either shamfully raised or shamefully deserted. So much is necessary that my old friends and pupils may understand my mind, and the repugnance I feel towards any pro-

ceeding which must record my connexion with University College. . . .
You will see then that I am altogether averse from lending aid or coun-
tenance to any scheme which will tend to remind others that I was a
teacher in the College which did homage to the evil it was created to
oppose.[90]

The rage eating away at Augustus De Morgan apparently weakened the ag-
ing mathematician physically as well as mentally. His wife later claimed that
it marked the beginning of his descent into mortal illness.[91]

Even in calmer moments of reflection, visions of ecumenical religious
harmony looked like mere fantasy by the end of De Morgan's life. When
James Martineau and J. J. Tayler, another nonconformist clergyman, wrote
in 1869 to numerous Londoners to gauge interest in establishing a new
communion (tentatively called the "Free Christian Union"), even the en-
thusiastic responses were tempered by skepticism. The Union would have
permitted all monotheistic believers—including Jews and Muslims, despite
the name of the organization—to set aside issues of dogma and gather together as a brotherhood under God. "I am interested in the attempt," De
Morgan excitedly wrote to Martineau, "which, hopeless as it seems to me to
the extent proposed, may yet originate a sect in which people may pray together without each man being fettered to his neighbours."[92] De Morgan approvingly told his wife that Christ and the Apostles would have joined the
Union.[93] However, De Morgan and others interested in the Union also worried about the sectarian extremism that permeated their society. These concerns surfaced in De Morgan's ambivalent reply to Tayler: "Supposing the
intermediates could fraternise with the extremes, could the extremes fraternise with one another? . . . You have your sand, and you aspire to make
rope."[94] How could anyone hope to bind the atomized laity together given
the charged and polarized religious atmosphere?[95]

John Venn (1834–1923), a follower of George Boole and creator of the
logical diagrams that bear his name, was one of many mathematicians who
realized that often each side of a fierce dispute has radically different assumptions. As he noted in *The Logic of Chance* (1866), for instance, those
who battled over the veracity of biblical miracles were separated by a chasm,
in which regardless of the format of the arguments—couched in mathematical terms or not—almost no chance for a resolution existed. "What is
to be complained of in so many popular discussions on the subject is the
entire absence of any recognition of the different ground on which the at-

tackers and defenders of miracles are so often really standing," Venn emphasized.[96] Materialists and others who scorn revealed religion simply have a viewpoint irreconcilable with those with faith in the Scripture, Venn argued: "How therefore can miraculous stories be . . . taken account of, when the disputants, on one side at least, are not prepared to admit their actual occurrence anywhere or at any time? How can any arrangement of bags and balls, or other mechanical or numerical illustrations of unlikely events, be admitted as fairly illustrative of miraculous occurrences?"[97] To Boole's ideal "laws of thought," universally found in every human mind, Venn therefore added a realist's understanding of human nature. We are stubborn, dogmatic, and indelibly colored by our assumptions and limited experiences; no amount of mathematical reasoning appears capable of closing the gap between those with starkly different beliefs.[98]

Resignation combined with a profound desire to escape from sectarian conflict, rather than ameliorate it, thus became an animating force in the era of mathematical professionalization. The distress mixtures of religion and mathematics caused among elite mathematicians in the 1860s and 1870s was indicative of a new cautious and possessive spirit that obscured the religious idealism of an earlier generation. Just as academic mathematicians tried to distinguish their practice from amateur mathematics, they also strove to distinguish it from theology.

A sharp-tongued spokesman for his vocation, Augustus De Morgan once again led the way. De Morgan abhorred and publicly derided mathematicians who supported religious dogma, as well as those who portrayed theological uses of mathematics as within the proper boundaries of the discipline. De Morgan collected examples of this kind of transgression, recounting and scathingly criticizing many of them on the pages of *The Athenaeum* in the 1860s and in his compilation *A Budget of Paradoxes*. He mercilessly denounced works such as Oliver Byrne's *The Creed of Saint Athanasius Proved by a Mathematical Parallel*, Olinthus Gregory's *Letter on the Christian Religion*, and especially Tresham Gregg's *Novum Organum Moralium*, with its use of his beloved mathematical logic.[99] (De Morgan sardonically pointed out that Gregg's Church, symbolized by the product $A\delta X$, would still be positive even with Apostles of "diabolical origin" $[-A]$ and "heretical doctrine" $[-\delta]$, since two negative quantities become positive when multiplied together.) Furthermore, De Morgan sought to make mathematical concepts less attractive to clergymen. For example, he shrewdly noted that illustrations using infinity to support Christian doctrine held true for other religions as well, including non-Christian and pagan creeds. Any dogma to which one grants "infinite" importance—no matter how absurd—will by definition require one's assent.[100] On the numerous theological uses of mathematics, De Morgan concluded, "The moral of all this is, that such things . . . should be kept out of the way of those who are not mathematicians, because they do not understand the argument; and of those who are, because they do."[101] In other words, theologians enamored of mathematics should leave the discipline to professionals—who, in turn, should never engage in theology.

De Morgan often responded to those who dared to use mathematics for theological purposes with sharp humor intended to undercut their legitimacy. One of De Morgan's favorite tales recalled how the mathematician Euler mocked the *philosophe* Diderot in front of Catherine the Great's court. As De Morgan delightedly retold it—not once but twice in *A Budget of Paradoxes*—Euler confronted Diderot, who was aggressively defending atheism to the great consternation of his audience. Euler claimed to have an incontrovertible mathematical demonstration of the existence of God that Diderot would have to explain or disprove. According to De Morgan, the sly Euler "advanced towards Diderot, and said gravely, and in a tone of perfect conviction: *Monsieur, (a + b^n)/n = x, donc Dieu existe; répondez!* Diderot, to whom algebra was Hebrew, was embarrassed and disconcerted; while peals of laughter rose on all sides. He asked permission to return to France at once, which was granted."[102] Thick dramatic embellishment betrays the passage's historical inauthenticity, of course, not to mention that Diderot was far from a novice in mathematics. Indeed, De Morgan's tale is actually an exaggerated version of unconfirmed gossip by a French courtier.[103] Even though the story of Euler and Diderot was a fiction, however, it was an important fiction for Augustus De Morgan, as well as for many professional mathematicians who recounted it in the late nineteenth and early twentieth centuries.[104] The myth encapsulated the absurdity of using mathematics for theological purposes, ridiculing people who did so as "philomaths" who deserved "a lesson in presumption."[105] Euler, though devout, knew (as every decent mathematician should) that it was ridiculous—even comical—to mix mathematics with religion.

A more potent force for distancing mathematics from faith than the derision of gadflies like De Morgan came from a broad push among professional mathematicians toward a nontranscendental philosophy of their discipline in the late Victorian era. In a way, this new philosophy reversed the logic of Tresham Gregg. If mathematics powerfully summoned divine ideas and symbols, it made sense that no prudent human being (including clergymen and amateurs) ought to ignore it; if mathematics was a human construct with practical utility but clear limitations, on the other hand, it safely could (and indeed should) remain in the hands of capable, secular professionals. By highlighting the discipline's earthly rather than heavenly origins, as well as its infirmities, mathematicians could make their field less alluring, and less threatening, to nonmathematicians.

One of Augustus De Morgan's students, William Stanley Jevons (1835–1882), anticipated this evolution of thought, eventually promoting distinct realms for mathematics and religion. Jevons began his career by formulating his own symbolic logic (*Pure Logic*, 1864), and he continued to work in the field as he carried its methods into economics. Initially he wished to extend George Boole's methodology and, like Boole, preserve a central role in society for religious knowledge and belief. Pure mathematics allows the mind to conceive and explore the infinite, as well as entities that do not conform to the normal laws of space and time, the Unitarian Jevons mused in his early studies. Such investigations inform the mind that the material realm might not be all that there is.

By the 1870s, however, Jevons had diverged from Boole's divine conception of mathematics, and more broadly, the transcendental conception of scientific knowledge. In his later thought appears a familiar modern compromise: science is powerful and worthy of respect, but ultimately limited in its purview and certainly not invalidating of religion, which functions in another sphere entirely. "The conclusions of scientific inference appear to be always of a hypothetical and provisional nature," he emphatically wrote in *The Principles of Science* (1874), "the best calculated results which it can give are never absolute probabilities; they are purely relative to the extent of our information."[106] Moreover, "Even mathematicians make statements which are not true with absolute generality."[107] Rather, Jevons claimed, mathematical and scientific knowledge—even knowledge aided by the clear terms and rigor of symbolic logic—could address only a limited realm and generally did so with mere probability rather than certainty. "The whole question now becomes one of probability and improbability," Jevons remarked about his new vision of scientific knowledge aided by mathematics, "We do not really leave the region of logic; we only leave that where certainty, affirmative or negative, is the result, and the agreement or disagreement of qualities the means of inference."[108] Science remained powerful and worthy of respect; at the same time, given its limitations, it could never destroy transcendental beliefs: "Atheism and materialism are no necessary results of scientific method. From the preceding reviews of the value of our scientific knowledge, I draw one distinct conclusion, that we cannot disprove the possibility of Divine interference in the course of nature. . . . From science, modestly pursued, with a due consciousness of the extreme finitude of our intellectual powers, there can arise only nobler and wider notions of the purpose of Creation."[109] As faith in scientific certainty waned, so would insidious and antireligious materialism, Jevons believed, because like theological uses of mathematics this secular philosophy drew strength from overly arrogant and expansive theories of scientific knowledge.

A fundamental shift in the understanding of mathematics originating on the Continent, away from the traditionally strong idealism of Anglo-American mathematicians, greatly strengthened feelings of uncertainty about the discipline. In large measure this transformation began with the insights of the French mathematician Augustin-Louis Cauchy (1787–1857), who realized in the early nineteenth century that both forms of the calculus sat on shaky foundations. Continental mathematicians had based their calculus on a number system lacking the imaginaries, and on a fundamentally unsound algebra. The calculus of Isaac Newton and his British followers, built on a flawed extension of geometry, was even more problematic. Disposing of these two versions, Cauchy worked to develop a comprehensive and rigorous version of the calculus, one that would include all numbers (real and imaginary) and that would have its basis in a more abstract notion of the limit than the physical sense derived from geometry. By investigating and publicizing the impurities of what had been regarded as "pure" mathematics, Cauchy set in motion a sweeping change in the way mathematicians understood their tools of trade.

In the Victorian age a flood of theoretical discoveries followed Cauchy's first wave of abstraction. In number theory, the branch of mathematics concerned with the integers, a series of central European mathematicians propelled the field far away from the enumeration of real things. A highly abstract understanding of numbers grew in the work of Karl Weierstrass (1815–1897), Richard Dedekind (1831–1916), Giuseppe Peano (1858–1932), and Georg Cantor (1845–1918).[110] Cantor also forged ahead in developing Boole's mathematical logic—in the process sapping it of any social relevance. For mathematical logicians after Cantor, the nature of the symbolical system itself took precedence over the coding and resolution of real-world arguments. The field became a pursuit of internal consistency, sharply divorced from the concerns of metaphysicians. As Cantor summarized the intention of his work in 1883, "Mathematics is entirely free in its development and its concepts are restricted only by the necessity of being non-contradictory and coordinated to concepts previously introduced by precise definitions. . . . The essence of mathematics lies in its freedom."[111] The Continental investigation of numbers and logic thus proffered a different meaning for the first word in "pure mathematics." This purity no longer implied a religiously tinged transcendentalism, but rather a conspicuous shunning of any relationship to the physical, metaphysical, or theological realms.

Perhaps the most striking example of such abstraction occurred in the field of geometry. Early in the Victorian period (and without much fanfare) two eastern European mathematicians, the Russian Nikolai Lobachevsky (1793–1856) and the Hungarian János Bolyai (1802–1860), challenged Euclid's fifth postulate. That time-honored law held that given a line and a point not on that line, one could draw only one parallel line through the point. By altering this assumption (which turns out not to be true on surfaces other than a flat plane) while maintaining Euclid's other postulates, Lobachevsky and Bolyai produced completely new forms of geometry. Moving further away from the ancient Greek mathematician's understanding of the world, in 1854 the German theorist Georg Riemann (1826–1866) established that non-Euclidean geometries could occur in spaces with more than three dimensions. A critical precursor to Einstein's general theory of relativity, Riemann's ideas severed geometry from common sense, transforming it into a province of esoteric abstraction. Geometry, which had once seemed so complete and independent of human experience, received a strong blow to its historical and philosophical foundations. As non-Euclidean geometry seeped into British mathematical thought in the last third of the nineteenth century, it began to trouble those who had pointed to geometry's supremacy as a science, a form of reasoning, and a handmaiden of natural theology.[112]

Victorian developments in several branches of mathematics combined to undermine many of the discipline's bedrock principles and methods, as well as its aura of transcendence. In field after field—the calculus, mathematical logic, geometry, algebra—the theoretical revolution made mathematics seem far from heavenly, a closed system of human axioms severed from reality and nature. Yet this was not an inexorable intellectual shift. So-

cial and professional concerns worked in concert with the changing philosophy of the discipline to weaken the formerly close relationship between
mathematics and faith, a movement particularly apparent in the views of
prominent late Victorian British mathematicians.

An influential promoter of the new philosophy of mathematics was
William Kingdon Clifford (1845–1879), who succeeded Augustus De Morgan at University College, London. When Clifford entered Trinity College,
Cambridge, in 1863, he was a staunch High Churchman who also read
deeply in Catholic theology, especially Thomas Aquinas.[113] In his academic
work, Clifford excelled at science in general and mathematics in particular,
showing a unique ability to visualize counterintuitive and complex topics.
He placed second in the Cambridge Tripos and became a Trinity fellow,
where he taught for two years as he began to publish advanced mathematical theories. Meanwhile, Clifford found himself deeply attracted to the ideas
of Charles Darwin and Herbert Spencer, leading to an intellectual transformation by the end of the 1860s. Where he had once used scientific analogies in the service of an orthodox theology, he now touted the power of a scientific method completely divorced from theological and metaphysical
concerns.[114] Clifford abandoned all interest in patristic texts and became
strongly anticlerical. He died before he could finish writing a book tentatively entitled *The Creed of Science*.[115]

Like De Morgan, Clifford used his position at UCL as a bully pulpit for
his mature views. He introduced Great Britain not only to the subject matter of non-Euclidean geometry (with his groundbreaking essay "On the
Space-Theory of Matter" in 1870), but also its profound philosophical implications. In his 1872 address "On the Aims and Instruments of Scientific
Thought," Clifford drew a distinction between "practically exact" and "theoretically exact" knowledge. Human beings attain the former through experience, meaning this knowledge could never be truly certain; the latter
offers complete precision and purity since it arises solely out of mental
processes. Most "truths" were only practically exact, Clifford thought, no
matter how sublime they appeared. Euclid believed he had attained theoretically exact knowledge in his postulates, which for two millennia seemed
to be unshakably, universally true. Had the earth been considerably smaller,
however—say, the size of a house—the Greek mathematician would have
realized that parallel lines can indeed intersect, just as lines of longitude
meet at the poles. Thus non-Euclidean geometry showed how seemingly ab-

solute mathematical laws were not transcendental after all, and pointed the way to a perhaps less revered, but more rigorous, set of axioms.[116]

Clifford also believed non-Euclidean geometry undercut those who made religious arguments using mathematical or scientific concepts. For instance, in 1875 Balfour Stewart (1828–1887), a physicist with a background in mathematics, and Peter Guthrie Tait (1831–1901), a mathematician with an interest in physics, published *The Unseen Universe; or Physical Speculations on a Future State*,[117] a book outlining how advanced theoretical physics supported the Christian view of the afterlife. Stewart and Tait published the work anonymously, but most of the scientific community knew who the authors were and eventually they added their names to a later edition.[118] The principle of the continuity (i.e., conservation) of energy in physics—energy cannot be destroyed, it can only change form—and its concept of the ether (an elusive substance forming the medium of space) strongly suggested, Stewart and Tait argued, that although the material human body may die, some part of us may live on as a form of energy in an invisible realm.

Although a colleague and friend of the two men, Clifford wrote a devastating critique of *The Unseen Universe* in *The Fortnightly Review*, which many like-minded scientists and mathematicians celebrated as a triumphant rebuke of theological pseudo-science. Clifford declared that Stewart and Tait, like Euclid, had made improper theoretical conclusions from a science that was only practically exact. How do we know that the laws of physics hold for a realm other than the only one we have access to—the visible, physical world? A concept such as "vortex-motion," the supposed movement of a continuous, frictionless liquid, was "a mere mathematical fiction" we can never apply with supreme certainty, Clifford insisted.[119] Molecules and the ether are only "complex mental images," he noted, "Is there anything that is not in our minds of which these things are pictures or symbols?" In other words, can we be sure that real external referents for Stewart's and Tait's fanciful conceptions exist? (Clifford had a good point here; twentieth-century physics dispelled the notion of the ether.) "In the last resort," Clifford firmly asserted, "all these questions of physical speculation abut upon a metaphysical question."[120] Mathematical laws of physics such as the inverse-square law of gravitation and Boyle's law of the density of gases are not necessarily "an external reality transcending phenomena," he emphasized. Like Euclid's axioms, they are married to a specific physical realm we have experienced, and rewritten in an idealized form. Therefore Clifford concluded that Stewart's and Tait's use of the law of continuity "has nothing to say to the question about the existence of something which is not matter, not phenomenon at all."[121] For William Kingdon Clifford, true scientists and mathematicians shunned inappropriate extrapolations such as this, and thus had no patience for the scientific support of theology.[122]

Although *The Unseen Universe* achieved great popularity and went through more than a dozen editions, Peter Guthrie Tait seems to have been chastened by attacks on his use of science to buttress Christian doctrine. A decade after the publication of the work (which in the intervening years had spawned many imitations[123]), a clergyman asked Tait to write an article proving how science leads its practitioners directly to the Christian faith. Tait briefly considered writing the piece, then turned down the offer. "While I see no objection to the occasional citation of such names as Faraday, Maxwell, Stokes, &c. for the purpose of showing that some of the very greatest of scientific men have been, and are, devout Christians; I think this should be done with caution," Tait wrote back, "For it was not their *science* that made them Christians; equally great scientific men have never been troubled themselves with the question. All that their cases prove is, that to their minds (minds trained to the highest attainable standard), there is *no contradiction* between science and any essential Christian verity."[124] *The Unseen Universe* sought to demonstrate how scientific laws could illuminate Christian dogma; now one of its authors retreated to the much weaker position that Christianity and science could get along. In subsequent public writing in the late 1880s, Tait backed away even further, merely suggesting that *religion* and science could coexist.[125] This timid compromise in the face of his colleagues' criticism allowed Tait to maintain his strong faith while setting his vocation apart from Christian apologetics.

Even for mathematicians and scientists who did not suppress their unalloyed support of faith, the advance of non-Euclidean geometry and other revolutionary concepts ultimately weakened their religious understanding of mathematics. Leading figures such as George Gabriel Stokes (1819–1903) began to revise their conceptions of the field. The son of a Church of Ireland rector, Stokes was the consummate Victorian mathematician, conversant in virtually all branches of the discipline as well as in associated topics in theoretical and experimental physics. He quickly bounded up the academic ladder, ultimately holding the Lucasian chair at Cambridge University for more than a half-century. Unlike many other contemporary mathematical luminaries, such as William Kingdon Clifford, Stokes underwent no major spiritual transformations during his lifetime. Although he eventually held more liberal ecclesiastical and theological opinions than his parents, Stokes sought to reform the Established Church from within. Moreover, he never felt the need to distance his scientific work from theology. For much of his life he actively participated in the Victoria Institute, a scientific body that strove to bolster natural theology.[126]

In a subtle way, however, Stokes's views were changing as the nineteenth century came to a close. For example, in 1890 he was asked to give the Gifford lectures at the University of Edinburgh. This series of lectures at four Scottish universities, endowed by Lord Adam Gifford in his 1885 will, focused on natural theology and its importance to intellectual life.[127] Unsurprisingly, Stokes accepted the offer. Privately, though, he was wracked by concerns. In a letter to Peter Guthrie Tait, Stokes described his consternation: "I feel appalled by the task I have undertaken. I could lecture on science with theistic bearings, or I could lecture on theology as a free lance, provided I were allowed to refer to revelation; but Lord Gifford has expressly excluded revelation, and wants his lecturers to build up a science of theology on a purely scientific foundation. . . . I confess I feel quite puzzled how possibly to fill 12 lectures, of presumably an hour's duration each, or thereabouts, with the theme prescribed by the founder as limited by his directions."[128] This trepidation caused Stokes to procrastinate. Nevertheless, he finally commenced in 1891 with an oration on divine suggestions in astronomy and physics, and in the following weeks Stokes spoke about chemistry, biology, and physiology.

As the series came to an end, however, the audience began to notice something odd. Strangely absent from the theological discussion was Stokes's very own specialty. His Gifford lectures contained no mathematics—not a single word on the topic. One of the foremost mathematicians of the day had spoken only about the physical and biological sciences in his defense of a rational belief in the existence of God.[129] Behind the scenes, the reason was clear. "I do not think that it is possible to prove the existence of God by mathematics," Stokes admitted to Tait in their 1890 correspondence.[130] Keenly aware of the ongoing reassessment of the foundations of mathematics, Stokes no longer associated the discipline with the lofty realm of theology. For him, God was to be found not in equations but in the evidence of the visible universe and the revelations of Scripture.

Rather than creating a protected sphere for the candid profession of religious convictions, the 1871 passage of the Tests Act, which opened the

Oxbridge schools to those outside of the Established Church, seems to have reinforced the growing hesitance among mathematics professors to include spiritual matters in their work and teaching. The life and thought of James Joseph Sylvester (1814–1897), a leading Victorian mathematician who as a Jew struggled his entire life to become part of the British establishment, plainly show this dynamic. A mathematical prodigy, Sylvester had the opportunity to engage many important theorists, including Augustus De Morgan, as a teenager. His religious background exposed him to abuse from classmates, however, and apparently after one such incident the boy ran away and sailed to Ireland. Returned home by an Irish judge, Sylvester eventually excelled once again in his academic work. He was able to study at St. John's College, Cambridge, for a time, yet could not take his degree because of the religious disqualification. (He finally received his Cambridge B.A. in 1872 at the age of 57.) The university also barred Sylvester from competing for the prestigious Smith's mathematical prizes, and prevented him from becoming a fellow at St. John's—once again due, in Sylvester's resentful words, to his "faith in which the Founder of Christianity was educated."[131] The young mathematician held a modest job at University College, London, for a few years, but despite being elected to the Royal Society by his mid-twenties Sylvester found it difficult to acquire a faculty position commensurate with his growing international fame. He spent four years teaching at the University of Virginia, ten years as an insurance actuary, and fifteen years at the Royal Military Academy in Woolwich before retiring in 1870, only a year before the colleges of Oxford and Cambridge finally opened their doors to non-Anglicans.

Soon after its founding in 1876, The Johns Hopkins University offered Sylvester a large salary and the opportunity to start a new department. The mathematician thought he had finally received an offer appropriate for his stature, and he once again left his native England to teach in the United States. Sylvester appears to have been reenergized by his American appointment, producing articles at a faster pace than ever before and establishing the *American Journal of Mathematics*. While overseas he also railed against the legacy of a policy that had kept him from a similarly high academic post in Britain. Sylvester scathingly declared in an 1877 address at Hopkins that the religious tests and "obscurantist class of [England's] university professors and heads" were to blame for making the country "so much inferior in intellectual weight and influence in the world." "Such is the blinding and blighting effect of early sectarian influences, one-sided culture, and narrow partisan connections," he concluded.[132] Nevertheless, in 1883, as he approached seventy, Oxford University offered Sylvester a coveted prize: the Savilian professorship of mathematics. He immediately accepted.[133]

Finally embraced by the British establishment, Sylvester promptly became mum on matters of religious belief, including the theological significance of mathematics. This reticence was a major shift from his prior attitude on these issues. As a student, Sylvester had studied under Olinthus Gregory at Woolwich and some of Gregory's grand notions about mathematics apparently rubbed off on his star pupil. In an 1854 lecture on geometry, Sylvester proclaimed that mathematics was "the language in which . . . the pages of the universe are written" and that "Sciences, true sciences, spring from celestial seeds sown in a mortal soil, they outgrow the restrictions which human shortsightedness seeks to impose upon them, and spread themselves outwards and upwards to the heavens from whence they derive their birth."[134] He toned down his language slightly in an address to the meeting of the British Association for the Advancement of Science in 1869, yet still marveled aloud about "the contemplation of divine beauty and order which [mathematics] induces, the harmonious connexion of its parts, the infinite hierarchy and absolute evidence of the truths with which it is concerned."[135] Throughout his life, Sylvester apparently sustained a belief that mathematics was "a divine philosophy," as he had called it in one of his early articles.[136]

Religious overtones disappeared from Sylvester's public statements, however, after his appointment to the Savilian professorship. In one of his first Oxford lectures in the mid-1880s, Sylvester portrayed his discipline as a thoroughly human endeavor, though an advanced and important one. He sought a golden mean between mathematics as quantitative trivia and problem solving, and mathematics as concepts and equations from a heavenly plane:

> An eminent colleague of mine, in a public lecture in the University
> . . . referred to a great mathematician as one who might possibly know
> every foot of distance between the earth and the moon; and when I
> was a member, at Woolwich, of the Government Committee of In-
> ventions, one of my colleagues, appealing to me to answer some ques-
> tion as to the number of cubic inches in a pipe, expressed his surprise

that I was not prepared with an immediate answer, and said he had supposed that I had all the tables of weights and measures at my fingers' ends. . . . I hope that in any class which I may have the pleasure of conducting in this University, other ideas will prevail as to the true scope of mathematical science as a branch of liberal learning . . . So long as we are content to be regarded as mere calculators we shall be the Pariahs of the University, living here on sufferance, instead of being regarded, as is our right and privilege, as the real leaders and pioneers of thought in it.[137]

Notably, his colleagues' ignorant comments did not impel Sylvester to pontificate about the cosmic significance of numbers and formulas. Rather, he merely tried to justify the place of mathematics (and himself) within the university in a statesmanlike way. Before his death, Sylvester founded the Oxford Mathematical Society to champion this view of mathematics as a distinct and progressive discipline, within the sphere of human knowledge but far above mere calculation.

Around the same time that James Joseph Sylvester assumed the Savilian chair at Oxford in the 1880s, the notion that mathematics was an incomprehensible yet important profession had begun to take hold among educated Britons. An 1885 review of one of William Kingdon Clifford's posthumous works in the newspaper *The Academy* illustrates how the public's divine sense of mathematics—and mathematicians—had devolved into a less mystical awe characterized by distant respect and resigned ignorance. "Mathematicians, compared with workers in other sciences and in the arts, are a secret body. What they do, how they do it, are unknown to the ordinary intelligence," the reviewer observed in a matter-of-fact tone. "Once upon a time, men who could compute the number of barleycorns required to go round the earth, or who could tell the multiplication table up to twenty-four times twenty-four or further, were called mathematicians; but now the name is found applied to men who do not eminently distinguish themselves in such calculations, and the world is puzzled."[138] The average person could comment upon the most recent efforts in literature and the arts, as well as upon scientific decrees such as the theory of evolution, the reviewer continued, "but the critic of the Darwinian theory, of the painting and the poem, has nothing to say of Taylor's theorem, and, because he knows no better, is quite content to live" without this knowledge.[139] In other words, no need to worry about mathematics. The discipline did not have anything to say about such lofty—and often theologically tinted—concerns as the origin of man or the nature of existence.

Charles Dodgson and Lewis Carroll

In an increasingly professional setting, late Victorian academic mathematicians who still wished to communicate to amateurs or sustain overt connections to religion often found themselves in a highly uncomfortable position. The career of Charles Lutwidge Dodgson (1832–1898), an Oxford don, clergyman, and mathematician who is better known for his children's fiction under the pseudonym Lewis Carroll, exhibits the difficulties imposed by the separation of professional mathematics from amateur mathematics and theology. Recent scholarship on Dodgson has focused on his sexual proclivities and social estrangement; he was an outcast in his first love—mathematics—as well.[140] In his attitude toward mathematical research, amateur mathematics, and the philosophy of the discipline, Dodgson violated all of the unwritten laws of the burgeoning mathematical profession. As a result, he found himself largely alienated and dejected as the nineteenth century came to a close.

After passing his examinations at Christ Church, Oxford, in 1852, Dodgson became a mathematics fellow there. In the succeeding decades, he produced a respectable fifty-eight publications on mathematics, over a half-dozen of them book length.[141] Moreover, although Dodgson used a pseudonym for his children's books, he used his real name for all of his early mathematical works, beginning in 1860 with *A Syllabus of Plane Algebraical Geometry, Systematically Arranged with Formal Definitions, Postulates, and Axioms*.[142] With that significant publication, he was well on his way to becoming an important figure in mathematical circles.

Unfortunately for his prospects, Dodgson yearned to be a popularizer of mathematics at the precise moment when other mathematicians were turning away from popular expressions and interpretations of their discipline. He failed to recognize the growing line between amateur and professional mathematics or to fathom the urge among his colleagues to thicken this boundary and join other academics in a self-serving aloofness. In his blunt 1876 poem "Fame's Penny Trumpet," Dodgson mocked "original researchers" as "little men of little souls." He entreated them to

Go, throng in each other's drawing rooms,
Ye idols of a petty clique:
Strut your brief hour in borrowed plumes
And make your penny-trumpet squeek.[143]

Never interested in the abstract mathematical forms that were undermining the ancient foundations of the discipline, Dodgson concentrated on Euclidean geometry in his independent work and in treatises intended for a general audience. As his colleagues pressed on to a new understanding of mathematics, Dodgson instead spoke of a "great cause which I have at heart—the vindication of Euclid's masterpiece."[144] In addition to his aversion to non-Euclidean geometry, Dodgson avoided the strange functions of symbolical algebra and four-dimensional mathematics. In short, he clung to the traditional idea that mathematics was a paradigm of simplicity and a conduit of absolute truth about the cosmos.[145]

Dodgson's best-known mathematical work, *Euclid and his Modern Rivals* (1879), desperately tried to vindicate an older style of mathematics against the new aesthetic. In stark contrast to his colleagues' journal articles and textbooks, Dodgson's book took the form of an entertaining play in which the characters described various facets of mathematics in plain language. He knew this tactic might lead to professional estrangement:

In one respect this book is an experiment, and may chance to prove a failure: I mean that I have not thought it necessary to maintain throughout the gravity of style which scientific writers usually affect, and which has somehow come to be regarded as an 'inseparable accident' of scientific teaching. . . . Pitying friends have warned me of the fate upon which I am rushing: they have predicted that, in thus abandoning the dignity of a scientific writer, I shall alienate the sympathies of all true scientific readers, who will regard the book as a mere *jeu d'esprit*, and will not trouble themselves to look for any serious argument in it.[146]

The content of *Euclid and his Modern Rivals* reflected Dodgson's view that mathematics had abandoned its once profound wisdom and relevance. He defended the primacy of Euclid's *Elements* in a somewhat quixotic attempt to stall the advance of the philosophical and pedagogical vanguard, which he believed was leading mathematics into a quagmire of esotericism and confusion. The criticism of Euclid's fifth postulate in non-Euclidean geometry was merely a diversion, Dodgson thought, which should have no bearing on the centrality of the ancient geometer to the discipline.

Furthermore, unlike most other professional mathematicians—and especially in contrast to Augustus De Morgan—Dodgson took a congenial, understanding tone toward circle squarers. For over a decade he corresponded with several of them.[147] Moreover, rather than calling them names or dismissing them outright as pesky cranks, he penned a humble work for circle squarers that spoke in calm and simple terms about why the endeavor was undeserving of their time and effort. Dodgson did not patronizingly speak of the problem in black and white, but rather prodded circle squarers to work through the matter for themselves.[148] He also empathized with circle squarers' bafflement about the strange conventions of modern mathematics. For instance, he agreed with John Parker about the peculiarity of lines having no width at all yet forming the basis of a reasonable and true discussion of reality. In an unpublished manuscript entitled "Pairs of Lines treated on Direction Theory," Dodgson accused his colleagues of improper abstraction: "So long as you do not assert the *existence* of such Lines, you are merely defining an imaginary concept; & your conclusions have no *practical* value, until you do assert their real existence."[149] In columns he wrote for the *Educational Times,* Dodgson responded to the concerns of many amateur mathematicians when he discussed the meaning of frustrating mathematical notions such as infinitesimals in plainspoken terms.[150] Ultimately finding himself nonplused by these concepts too, he derided the "bewildering region of Infinities and Infinitesimals" in his 1888 book *Curiosa Mathematica.*[151]

In addition, Dodgson adopted a transcendental view of mathematics that matched the vision of an earlier generation but which increasingly fell out of favor with his contemporaries. He spoke of mathematical geniuses scaling great mountains in their work, eventually moving "so near the heavens."[152] Worse, Dodgson had the bad luck to be educated and ordained as a deacon in an era when clerical mathematicians were common, but reached middle age just as the two professions separated. Between 1831 and the founding of the London Mathematical Society in 1865, fifteen Anglican clergymen served as president of the mathematical section of the British Association for the Advancement of Science; in the next thirty-five years, only two clerics held that position.[153] By 1884, fewer than one in ten members of the London Mathematical Society were clergymen.[154] Despite these dwindling numbers, Charles Dodgson continued to believe not only in the possibility of clergymen participating broadly in the mathematical profession, but also in the benefits of this participation. Recounting his young adulthood, when he was ordained and became a mathematics lecturer at Oxford, Dodgson noted how he "came to the conclusion that, so far from educational work (even Mathematics) being [an] unfit occupation for a clergyman, it was distinctly a *good* thing that many of our educators should be men in Holy Orders."[155] In 1885, Dodgson recommended Lady Margaret Hall, Oxford, to a girl who wished to study mathematics because he felt it provided a sound and traditional mathematical education while being "conducted on Church principles."[156]

But by the 1880s the intellectual and institutional environment of Dodgson's youth had changed around him. Despite his appointment at Christ Church being a clear first step to a career in mathematics, Dodgson never felt comfortable enough to join any of the mathematical societies that sprang up in the last third of the century. He tended to work alone and often failed to attribute sources he used in his proofs.[157] Eventually finding himself too far outside of the elite circle of mathematics, Dodgson resigned his lectureship in 1881. Four years later he published a mathematical work under his pseudonym, Lewis Carroll, for the first time. Whereas he titled somber earlier books prosaically, such as *An Elementary Treatise on Determinants with Their Application to Simultaneous Linear Equations and Algebraical Geometry* (1867),[158] he gave his new book a fanciful and mysterious title: *A Tangled Tale* (1885). This compilation of mathematical puzzles and riddles, originally published as a series in the popular Church of England circular *The Monthly Packet*, returned to the ideals of the *Gentleman's* and *Lady's Diaries*. In each puzzle, Dodgson wrote in the preface to *A Tangled Tale*, he was trying to "embody . . . one or more mathematical questions—in Arithmetic, Algebra, or Geometry, as the case might be—for the amusement, and possible edification" of the layperson.[159]

His remaining works in mathematics and related fields alternated between attribution to Charles Dodgson and Lewis Carroll, yet the jocose Carroll had clearly come to the fore. In the preface to *The Game of Logic* (1887), published under his pseudonym, Dodgson spoke of the topic as providing "a little instruction" but ensured the reader it was also "an endless source of amusement."[160] He bordered on comical histrionics in his *Symbolic Logic* (1896; also published as Lewis Carroll), exclaiming that the subject provided "one of the most, if not *the* most, fascinating of mental recreations!"[161] Even books published under the name Charles Dodgson, such as *Curiosa Mathematica,* had a playful character that set them apart from the serious tone and research agenda of mathematicians like John Venn. Unable and unwilling to adapt to a professional culture that had eschewed the concerns of amateurs and the clergy, Dodgson receded deeper and deeper into his alter ego.[162]

Pure Mathematics at the Turn of the Century

The attitude and philosophy that alienated Charles Dodgson hardened in the generation of leading British and American mathematicians who succeeded Clifford, Stokes, and Sylvester. Many in this *fin de siècle* generation ventured in the 1880s to German universities such as Göttingen and Leipzig—epicenters of the new math.[163] They sought out theorists like Felix Klein, one of the foremost researchers in non-Euclidean geometry, whose Göttingen program (and later his program at Johns Hopkins) married an earthly vision of mathematics with an elitist sensibility.[164] Klein's model became the archetype for new research sites such as the University of Chicago, as well as recast programs at Oxford, Cambridge, Harvard, Princeton, and Yale. Expanding on this German influence, Anglo-American mathematicians developed an even firmer sense of professionalism while adding to their discipline's esoteric abstraction. The American George Bruce Halsted supplied the first English translations of Lobachevsky's and Bolyai's works on non-Euclidean geometry in the 1890s, leading to widespread interest in the topic.[165] In Britain, Alfred North Whitehead's groundbreaking work *Universal Algebra* (1898) forever severed the age-old ties between algebra and arithmetic. "It is obvious that we can take any marks we like and manipulate them according to any rule we choose to assign," Whitehead told his colleagues about algebraic notation.[166]

The cultural and intellectual forces swirling around mathematics coalesced most profoundly in Whitehead's colleague, and occasional coauthor, Bertrand Russell (1872–1970). While many historians and philosophers have closely examined the development of Russell's early thought, the focus seems justified—perhaps more than any other figure, the young Russell displayed the intellectual confusions and transitions of the late nineteenth and early twentieth centuries.[167] After losing his parents at a young age, he set out on a turbulent spiritual journey marked by the influence of John Stuart Mill and William Kingdon Clifford, the idealists F. H. Bradley and J. M. E. McTaggart, and pure mathematics and logic from the Continent.

Russell inherited his family's contentiousness and exhibited it from an early age, often in petulant objections to rote lessons and ideological assumptions. When Russell first tried his hand at geometry and algebra in the early 1880s, he found the fields' use of unproven axioms distinctly upsetting.[168] Similarly, he questioned the dogma and practice of the Church of England. Russell's grandmother, who wavered between orthodoxy and nonconformity (she ended her life as a Unitarian), forced him to attend church and read the Scriptures regularly. However, like his father and brother before him, Russell eventually reached a state of extreme religious doubt, casting off the authority of the Bible and drifting toward a strongly agnostic position. Herbert Spencer, a close friend of Russell's grandmother, probably had some effect, as did his reading of Charles Darwin; perhaps more directly important, however, was William Kingdon Clifford, whom Russell read intently in the late 1880s. Clifford's simultaneous praise of science and criticism of those who believed they held "theoretically exact" knowledge struck a balance that seemed just about right to the maturing, increasingly antiauthoritarian Russell.[169] Also contributing to Russell's growing skepticism was his godfather John Stuart Mill's autobiographical description of his own spiritual tumult.[170] Matters came to a head in 1890. Russell wrote in his diary that he was "in such a haze about fundamental axioms" that he was "lean[ing] towards the opinion that nothing is to be accepted which is incapable of experimental or inductive proof."[171] He was cloudier than ever about his beliefs when he matriculated to Cambridge University in his late teens.

At Cambridge, Russell found some measure of solace in the idealist philosophy of F. H. Bradley. Bradley, one of the premier British metaphysicians of the age, advocated a brand of idealism that emphasized an individual's intuition of the cosmos—"the immediate unity which comes in feeling," as he called it.[172] Russell's undergraduate essays clearly exhibit the comforting influence of Bradley. For example, in an 1894 paper written for James Ward's undergraduate course on metaphysics, Russell discussed various mental states, and he highlighted one he believed played an important role in discovering "meaning": "If I lie in a field on a hot day with my eyes shut, and feel sleepily the heat of the sun, the buzz of the flies, the slight tickling of a few blades of grass, it is possible to get into a frame of mind which seems to belong to a much earlier stage of evolution; at such times there is only what Bradley would call 'a vague mass of the felt.'"[173] Under the sway of Bradleyan idealism, the young Russell experimented with a wide range of philosophical modes, and managed to keep atheism at bay.

In 1895, Russell scored well on the Cambridge Moral Sciences Tripos and became a fellow at Trinity College, the beginning of a long and highly successful academic career. However, his interest in the cosmic notions and "feeling" of F. H. Bradley was beginning to wane under the influence of another, less transcendental idealist: J. M. E. McTaggart, an unorthodox Hegelian who maintained a highly *un*emotional view of philosophy. McTaggart summarized his philosophical aesthetic when he wrote that although we need both feeling and analysis, "I don't think it's any good appealing . . . to the heart on questions of truth. After all there is only one way of getting at the truth and that is by proving it."[174] This hard-nosed approach rekindled Russell's earlier dissatisfaction with assumptions in all areas of intellectual life. Under McTaggart's sway, Russell came to the conclusion that the metaphysical intuitions of Bradley were not legitimate sources of knowledge, and in a lecture at Cambridge in 1899 he publicly abandoned his attachment to his former mentor's idealism. Within a few years he had come to believe in two main impulses in the human mind, "mysticism and logic," which, although both necessary to human existence, were diametrically opposed. Any true and rigorous philosophy must endeavor to exist on only one side of this dichotomy, Russell thought, and he now chose the logical side over the mystical.[175] While Russell's new ideology did not banish religion from the world, it pushed it aside: religion should *complement* academic concerns, not *inform* them.

Russell's studies in a discipline that had once frustrated him, pure mathematics, contributed to this major philosophical shift. One of his first published works in the field, *An Essay on the Foundations of Geometry* (1897), showed the profound influence of non-Euclidean geometry on his thought. Russell noted in the introduction that geometry was once the "impregnable fortress of the idealists," an exemplar of knowledge that most human beings considered "independent of experience" and universally objective.[176] In the late nineteenth century, however, that "cherished religion of our childhood" crumbled as mathematicians lost faith in Euclid's axioms.[177] As with William Kingdon Clifford, therefore, non-Euclidean geometry appears to have erased any vestiges of religious idealism from Russell's mind. The unease he felt with geometry as a child grew into an indictment of all pretensions to "theoretically exact" knowledge. Russell's shattered trust in Euclid's *Elements* became a cautionary tale about authority and certainty, its moral applicable far beyond the realm of mathematics. Subsequently, Russell eagerly explored other advances in pure mathematics in the final years of the nineteenth century, especially the logical work of Giuseppe Peano and Gottlob Frege. In addition, like many of his colleagues Russell spent time in Germany studying with cutting-edge mathematical theorists.

By 1900, Russell had found his mission: to complete the redefinition of mathematics, to divorce it once and for all from the real world and from numbers. He succeeded by extending the insights of symbolic logic until he proved that mathematics and logic were at heart the same thing—you could construct mathematics out of certain basic laws of logic. This epiphany resulted from a purely mental exercise, one completely opposite from the pure thought hailed in the discovery of Neptune. Whereas the minds of Urbain Le Verrier and John Couch Adams supposedly opened heavenward to reveal the elegant congruency between mathematical laws and the composition of the universe, Russell's mathematical contemplation folded, like origami, onto itself. Unlike Isaac Newton's *Principia Mathematica*, Bertrand Russell's *The Principles of Mathematics* (1903) simply described mathematics, with no implications for understanding how the cosmos functioned.

Two years before publishing that seminal volume, Russell wrote an article for the popular audience of the *International Monthly* to explain late-nineteenth-century advances in mathematics and the significance of his own work. In gleeful terms, Russell began the article by claiming that "the nineteenth century, which prided itself upon the invention of steam and evolution, might have derived a more legitimate title to fame from the discovery of pure mathematics."[178] He marveled at giants of recent logic and mathematics, such as Weierstrass, Dedekind, and Cantor: "I know of no age (except perhaps the golden age of Greece) which has a more convincing proof to offer of the transcendent genius of its great men."[179] Yet Russell's praise of these pure mathematicians only superficially mimicked early Victorian odes. What made these theorists so awe inspiring, he argued, was not that they unveiled God's thoughts, but that they gave us a much better understanding of the human mind and the concepts we use without reflection. Russell credited George Boole's *The Laws of Thought* with sparking this revolution, since (according to Russell) the technique it presented did not care if the propositions replaced by mathematical symbols were actually true, or even if they represented anything at all.[180] Giuseppe Peano further perfected Boole's "strict symbolic form," devoid of relevance to anything other than itself, by removing all words from the system.[181]

Furthermore, Russell explained, once-revered mathematical notions now disclosed their true nature to researchers. Mathematicians had exposed the grand and theologically pregnant sense of infinity as the *res summa*—that which is greater than everything else—as merely a "specious maxim." Instead, Cantor and Dedekind offered a counterintuitive, "harmless definition of infinity," which described it as a collection of terms that "contains as parts other collections which have just as many terms as it has." Worse still for the common understanding of infinity, Russell noted that pure mathematicians had uncovered not just one, but a host of infinities.[182] Similarly, geometry, formerly "the study of the space in which we live," had changed its stripes in the new philosophy of mathematics. Russell recounted the impact of non-Euclidean geometry in a way that would have elated William Kingdon Clifford and exasperated Charles Dodgson: "It has gradually appeared, by the increase of non-Euclidean systems, that Geometry throws no more light upon the nature of space than Arithmetic throws upon the population of the United States. . . . Whether Euclid's axioms are true, is a question as to which the pure mathematician is indifferent; and what is more, it is a question which it is theoretically impossible to answer with certainty in the affirmative. . . . The [modern] geometer takes any set of axioms that seem interesting, and deduces their consequences."[183] To further distinguish this new sense of geometry from the traditional one, Russell emphasized that "in the best books [on geometry] there are no figures at all."[184] So much for the once-beloved Euclid and his *Elements;* with noticeable relish, Russell impugned Euclid as a mere historical curiosity, still popular but entirely irrelevant for the future of mathematical research or education.[185]

By trivializing well-known mathematical texts such as Euclid's *Elements,* Russell drew a strict line between popular and elite conceptions of mathematics. He mocked rudimentary methods, like counting (which many amateurs still considered the totality of arithmetic), as a "very vulgar and elementary way of finding out how many terms there are in a collection."[186] Moreover, Russell understood that the advances he energetically advocated would come as "rude shocks to mathematical faith." To those outside the academy, the new philosophy of mathematics would no doubt appear as "outrageous pedantry," just as the irrationality of π had made circle squarers livid and antagonistic toward mathematics professors.[187] Too bad, Russell told his audience—mathematicians had an obligation to push formalism to its rightful end. Besides, he noted, those who had always found mathematics perplexing might "find comfort" in the notion that mathematicians dabbled in ideas thoroughly unrelated to anything outside of the discipline.[188]

"Mathematics may be defined as the subject in which we never know what we are talking about, nor whether what we are saying is true," Bertrand Russell thus concluded in the first year of the twentieth century.[189] It is impossible to imagine a greater contradiction of the deeply held faith of Galileo, Kepler, and Newton, John Herschel and Benjamin Peirce. An unsettling statement even a hundred years later, Russell's public definition of mathematics was the logical culmination of a half-century of mathematicians' distancing their subject matter from physical reality and the ideals of amateurs and theologians. Mathematics, once loudly declared by its practitioners to be the pursuit of God's equations, entered the twentieth century in a far more modest guise.

Notes

INTRODUCTION The Allure of Pure Mathematics in the Victorian Age

1. For a complete history of this planetary discovery, see Grosser, *The Discovery of Neptune* and Standage, *The Neptune File*.

2. Robert Harry Inglis, "Address," in *Report of the Seventeenth Meeting of the British Association for the Advancement of Science* (London: John Murray, 1848), xxx. Emphasis in the original.

3. John Herschel, "An Address Delivered at the Annual General Meeting of the Royal Astronomical Society, February 11, 1848, on the Subject of the Award of the Testimonials," in *Memoirs of the Royal Astronomical Society* (London: Royal Astronomical Society, 1849), 17:173. Emphasis in the original.

4. S. De Morgan, *Augustus De Morgan* 129.

5. Herschel, "An Address," 173–174.

6. Quoted in Grosser, *The Discovery of Neptune* 142.

7. Brewster, "Researches Respecting the New Planet Neptune," 111.

8. Nichol, *The Planet Neptune* 83.

9. William Rowan Hamilton to Maria Edgeworth, 8 February 1847, in Graves, *Life of Sir William Rowan Hamilton* 2:552. Letters to and from Hamilton, reprinted in their entirety in this work, have been checked with the originals when possible.

10. Ibid. Emphasis in the original.

11. S. De Morgan, *Augustus De Morgan* 131.

12. Bartol, "The New Planet," 77.

13. See Bode, *Anleitung zur Kenntniss des gestirnen Himmels*. Bode was a Swiss naturalist who was fascinated (like many other eighteenth-century intellectuals, including Immanuel Kant) by the possibility of a harmonic description of the planets. He found his mathematical "law" in the work of Johann Daniel Titius. See Grosser, *The Discovery of Neptune* 27–29.

14. Peirce, *Ideality in the Physical Sciences* 12–13.

15. Nichol, *The Planet Neptune* 9–10.

16. Ibid., 75.

17. Somerville, *Personal Recollections* 140–141.

18. Christmas, *Echoes of the Universe* 131.

19. Ibid., 132.

20. Bushnell, *An Oration Delivered Before the Society of Phi Beta Kappa* 32.

21. Ibid.

22. Ibid., 31.

23. Ibid., 32.

24. Farrar, *Sermons Preached in St. Mary's, Oxford* 47.

25. Ibid.

26. Ibid., 48.

27. Everett, *Orations and Speeches* 3:514.

28. Boole, *Studies in Logic and Probability* 195.

29. Most of the works on the history of symbolic logic are predominantly outlines of its technical development. See, e.g., Styazhkin, *History of Mathematical Logic;* Merrill, *Augustus De Morgan and the Logic of Relations;* and Shearman, *The Development of Symbolic Logic.*

30. For this historiographical perspective, see, most famously, Russell, *A History of Western Philosophy.*

31. Unfortunately, most histories of mathematics are far more parochial than the topic demands, such as Boyer's classic study *A History of Mathematics.* Many historical studies have been written by mathematicians themselves, providing serviceable technical accounts of their discipline without addressing the wider cultural and historiographical issues that are of interest to mainstream historians. Only recently have some historians attempted to integrate pure mathematics into a larger understanding of European and American history and culture, though one suspects that they have only grazed the surface of the matter. See Rowe and McCleary, *The History of Modern Mathematics;* Mehrtens et al., *Social History of Nineteenth Century Mathematics;* and Parshall and Rowe, *The Emergence of the American Mathematical Research Community, 1876–1900.* For ideas most germane to this study, see the work of Joan L. Richards, especially *Mathematical Visions* and "God, Truth, and Mathematics in Nineteenth Century England." Some of the themes of this study—e.g., the tension between certainty and contingency in mathematics—also come out vividly in Richards's remarkable combination of autobiography and research into the life of Augustus De Morgan in *Angles of Reflection.*

32. For a discussion of this confusion in the realm of mathematical physics, see Olson, *Scottish Philosophy and British Physics, 1750–1880* 188–193.

CHAPTER ONE Heavenly Symbols

1. On Whewell's significance as a revered formulator of the philosophy of science, see Yeo, *Defining Science.*

2. For a contemporary's view of Whewell and his work in mathematics, see Todhunter, *William Whewell: An Account of His Writings.*

3. Whewell, *The Philosophy of the Inductive Sciences Founded Upon Their History* 1:153–154.

4. Ibid., 2:122–128.

5. Ibid., 139. The quotation is from *Historia Naturalis* I.75.

6. Ibid., 156–160.

7. Herschel, *A Preliminary Discourse* 104–117.

8. See Smith, *The Boole-De Morgan Correspondence 1842–1864* 77; Hill, *Geometry and Faith*, 2d ed., frontispiece.

9. William Rowan Hamilton to Augustus De Morgan, 8 December 1851, Ms 1493/494, Hamilton Papers, Special Collections, Trinity College Library, Trinity College, Dublin; Graves, *William Rowan Hamilton* 2:525.

10. Plato, *The Republic* 177 (525a–b in the standardized Stephanus enumeration).

11. Ibid., 177 (525c).

12. Ibid., 178–179 (527b–c).

13. Ibid., 164–166 (510–511).

14. Ibid., 187 (536d). Emphasis is mine.

15. Plato, *Timaeus* 25 (35b–36b).

16. Ibid., 55–59 (53c–55c).

17. In this respect, as detailed later, they extended a long tradition of treating the *Timaeus* as one of Plato's most important works. The *Timaeus*'s stock would fall sharply during the nineteenth century, however, and eventually become relegated to the margins of Plato scholarship.

18. A. C. Lloyd, introduction to the *Epinomis*, in Plato, *Philebus and Epinomis* 205.

19. Ibid., 221 (973b).

20. Ibid., 229 (978c). As Galileo later famously wrote, "Philosophy is written in this grand book, the universe, which stands continually open to our gaze. But the book cannot be understood unless one first learns to comprehend the language and read the letters in which it was composed. It is written in the language of mathematics." Galilei, "The Assayer," in *Discoveries and Opinions of Galileo* 237–238.

21. Ibid., 227 (977b).

22. Ibid., 228 (977d–e).

23. Ibid., 227 (977c).

24. Lloyd, introduction to the *Epinomis*, 211–217.

25. Plato, *Epinomis*, 222 (973c).

26. Proclus, *Philosophical and Mathematical Commentaries*. On Taylor's own work and philosophy, see Taylor, *Thomas Taylor the Platonist: Selected Writings*. Taylor had provided the first English translation of the *Epinomis* in 1804. See Plato, *Opera*.

27. This description of mathematics was echoed by the other major post-Plotinian Neoplatonist, Iamblichus. However, Iamblichus's work *De Communi Mathematica Scientia* was not as well known in the English-speaking world as Proclus's translated works. For a technical discussion of the mediation of mathematics in the work of both Iamblichus and Proclus, see Merlan, *From Platonism to Neoplatonism* 11–33. For more on the Platonic sources of mathematics-as-intermediary, see Jacob Klein, *Greek Mathematical Thought* 69–79.

28. Proclus, *Commentaries* 3.

29. Ibid., 4.

30. Ibid.

31. Ibid., 38.

32. Ibid.

33. Ibid., 4.

34. R. T. Wallis, *Neoplatonism* 170.

35. Ibid., 173.

36. Ficino, *Three Books on Life* 115.

37. George Boole to Augustus De Morgan, 21 March 1859, in Smith, *The Boole-De Morgan Correspondence* 77.

38. John Dee, preface to Euclid, *Elements* 2–3.

39. Ibid., 3.

40. Peirce, *Linear Associative Algebra* 1 (page citations are to the reprint edition).

41. Dee, preface to Euclid, *Elements* 3–4.

42. For a broader account of Norris, see Muirhead, *The Platonic Tradition in Anglo-Saxon Philosophy* chapters 4 and 5.

43. Norris, *An Essay* 1:402.

44. Ibid., 59–60. Emphasis in the original.

45. Ibid., 372.

46. For a more technical discussion of Wallis's theory of mathematics, see Scott, *The Mathematical Work of John Wallis* and Klein, *Greek Mathematical Thought* 211–224.

47. Augustus De Morgan to George Boole, 16 October 1861, Ms Add. 97, Special Collections, University College Library, University College, London.

48. Klein, *Greek Mathematical Thought* 218.

49. John Wallis, *Treatise of Algebra* 273.

50. *Encyclopaedia Britannica*, 9th ed., 24:332.

51. For a broader treatment of Llull, see Johnston, *Spiritual Logic*.

52. Styazhkin, *History of Mathematical Logic* 10–11.

53. Quoted in Prantl, *Geschichte der Logik im Abendlande* 3:150; trans. in Styazhkin, *History of Mathematical Logic* 11.

54. Styazhkin, *History of Mathematical Logic* 12.

55. Yates, *The Art of Memory* 173ff.

56. Peter J. French, *John Dee* 44–47.

57. See Seaton, "Thomas Harriot's Secret Script," 111–114.

58. Shirley, *Thomas Harriot* 109–112.

59. Augustus De Morgan to John Herschel, n.d., Ms HS.6.356, Herschel Papers, Royal Society of London, London.

60. Wilkins, *Mathematical and Philosophical Works* 2:247.

61. Ibid., 249.

62. Ibid., 255.

63. Ibid., 248.

64. Ibid.

65. Herschel, *Natural Philosophy* 19–20.

66. Ibid.

67. William Wordsworth, *Excursion,* quoted in Inge, *Platonic Tradition* 71.

68. For a more technical discussion of Kant's own views on the nature of mathematics, see Friedman, "Kant's Theory of Geometry," 455–506.

69. Kant, *Critique of Pure Reason* 587.

70. In his formative years Coleridge read the classic Neoplatonists Iamblichus, Proclus, Porphyry, and Plotinus. See Samuel Taylor Coleridge to John Thelwall, 19 November 1796, in Coleridge, *Collected Letters* 1:262. For a discussion of Coleridge and early modern English Neoplatonism, see Schrickx, "Coleridge and the Cambridge Platonists," 71–91.

71. See the section on Boole's *The Laws of Thought* in chapter 3. On Coleridge and the dilemmas posed by Spinoza's thought, see McFarland, *Coleridge and the Pantheist Tradition.*

72. Coleridge, *Biographia Literaria* 1:242.

73. Coleridge, *Aids to Reflection* 388–389.

74. Ibid., 389.

75. Ibid., 390.

76. Coleridge, *Biographia Literaria,* 1:133. Emphasis in the original.

77. Ibid., 134–135. Emphasis in the original.

78. For a general discussion of Coleridge's philosophy in relation to the natural sciences, see Levere, "S. T. Coleridge: A Poet's View of Science," 34–44.

79. Quoted in Levere, "S. T. Coleridge: A Poet's View of Science," 37. Emphasis in the original.

80. Coleridge, *Collected Works* 4:478.

81. Trevor H. Levere, "Coleridge, Chemistry, and the Philosophy of Nature," *Studies in Romanticism* 16 (1977): 375.

82. Levere, "S. T. Coleridge: A Poet's View of Science," 35.

83. Coleridge, *Notebooks* vol. 2, *1804–1808,* entry 2894. Emphasis in the original.

84. In the British Museum copy of that work, Coleridge noted Platonically that the correct path of education for a philosopher would include "elements of Geometry and universal Arithmetic." Quoted in J. R. de J. Jackson, introduction to Coleridge, *Logic* lx.

85. Coleridge, *Biographia Literaria* 1:156. Emphasis in the original.

86. Coleridge, *Notebooks* vol. 2, *1804–1808,* entry 2546.

87. Coleridge, *Aids to Reflection* xi.

88. Samuel Taylor Coleridge to C. A. Tulk, 20 January 1820, in Coleridge, *Collected Letters* 5:19.

89. Coleridge, *Biographia Literaria* 1:203.

90. The completion date of the manuscript is uncertain; see J. R. de J. Jackson, introduction to Coleridge, *Logic* xxxix–li. On Coleridge's logic, see Snyder, *Coleridge on Logic and Learning.*

91. For Coleridge's debt to Kant, see McFarland, *Coleridge and the Pantheist Tradition*, and Orsini, *Coleridge and German Idealism*.

92. Coleridge, *Logic* 201–202.

93. Ibid., 199.

94. Ibid.

95. Akenside, "The Pleasures of Imagination," in *Poems* 156–157; quoted in Coleridge, *Logic* 199.

96. Wordsworth, *The Prelude* 89 (page citations are to the reprint edition).

97. Ibid.

98. Quoted in Inge, *Platonic Tradition* 73.

99. William Rowan Hamilton to Lord Adare, 19 April 1842, Ms 1492/57a, Hamilton Papers, Special Collections, Trinity College Library, Trinity College, Dublin.

100. Ms 7762–72/2089, Hamilton Papers, Special Collections, Trinity College Library, Trinity College, Dublin.

101. See Hankins, *Sir William Rowan Hamilton* 103.

102. Graves, *William Rowan Hamilton* 1:193–194.

103. Nichol, *The Planet Neptune* 90.

104. On the British religious setting, see Chadwick, *The Victorian Church;* Welch, *Protestant Thought in the Nineteenth Century;* Reardon, *From Coleridge to Gore;* Prickett, *Romanticism and Religion.* On the American religious setting, see the later chapters of Jon Butler, *Awash in a Sea of Faith,* and Hatch, *The Democratization of American Christianity.*

105. Mary Everest Boole, "Home-side of a Scientific Mind," in *Collected Works* 1:12. Hereafter referred to as "Home-side."

106. Ibid., 39.

107. Boole, "The Fellowship of the Dead," Ms 12.K.45(ii), Special Collections, Royal Irish Academy, Dublin.

108. Ibid.

109. Ibid.

110. Ibid.

CHAPTER TWO God and Math at Harvard

1. Raymond Clare Archibald adopted the phrase for his introductory essay as editor of *Benjamin Peirce: 1809–1880.*

2. Peabody, *Elegy at the Funeral of Benjamin Peirce* 1.

3. Byerly, "Reminiscences of Peirce," in Archibald, *Benjamin Peirce: 1809–1880* 5–6.

4. No full-length biographies of Peirce exist. Two pamphlet-sized sketches of Peirce are Rantoul's *Memoir of Benjamin Peirce* and Archibald's introductory essay to *Benjamin Peirce: 1809–1880.* For a shorter sketch outlining Peirce's major contributions to mathematics and science see Hill, "Benjamin Peirce," 91–92.

5. Rantoul, *Memoir of Benjamin Peirce* 3–4.

6. Ahlstrom, *A Religious History of the American People* 398.

7. Clarke, *Autobiography, Diary, and Correspondence* 34.

8. Peabody, *Harvard Reminiscences* 181. This was a fairly impressive feat. According to Florian Cajori, the normal mathematical curriculum at Harvard in about 1830 consisted of algebra and solid geometry (freshman year), trigonometry, topography, and the calculus (sophomore year), and topics in natural philosophy (mechanics, optics, and so on; senior year). See Cajori, *The Teaching and History of Mathematics in the United States* 133.

9. Emerson, "The Transcendentalist," in *The Complete Essays* 93. See Richardson, *Emerson: The Mind on Fire* 250–251.

10. For the pre-nineteenth-century history of Unitarianism in the United States, see Wright, *The Beginnings of Unitarianism in America*. For Britain, see Richey, *Origins of English Unitarianism*.

11. Wright, *The Liberal Christians* 36.

12. Quoted in Weiss, *Discourse Occasioned by the Death of Convers Francis, D.D.* 28–29.

13. Ladu, "Channing and Transcendentalism," 136–137.

14. On Parker and Transcendentalism, see Dirks, *The Critical Theology of Theodore Parker*; Smith, "Was Theodore Parker a Transcendentalist?" 1–32; and Wright, *The Liberal Christians*, 34–35.

15. Parker, *The Transient and Permanent in Christianity* 6.

16. Ibid.

17. Ibid., 9.

18. Ibid., 8.

19. Ibid., 12.

20. Ibid., 18.

21. Ibid., 30.

22. Charles Sanders Peirce, *Collected Papers* 6:86–87.

23. Emerson, *The Early Years of the Saturday Club* 2.

24. Emerson, "The Transcendentalist," 87.

25. Emerson, "An Address," in *The Complete Essays* 80.

26. Ibid., 72.

27. Ibid., 75.

28. Ibid., 79–80. Emphasis in the original.

29. Ibid., 75.

30. Ibid., 80.

31. Ibid., 68.

32. Ibid., 84.

33. Emerson, "The Transcendentalist," 93–98.

34. Emerson, "Nature," in *The Collected Works of Ralph Waldo Emerson* 1:33. On Emerson's reverence for Plato, see Emerson, "Plato; or, the Philosopher," in *The Complete Essays* 471–498.

35. Ibid., 34.

36. Ibid.

37. Emerson seems to have been frustrated by the subject in college. His journal entry of 15 October 1820 records: "My more fortunate neighbours exult in the display of mathematical study, while I after feeling the humiliating sense of dependence & inferiority which like the goading soul-sickening sense of extreme poverty, palsies effort, esteem myself abundantly compensated, if with my pen, I can marshal whole catalogues of nouns & verbs, to express to the life the imbecility I felt." *Emerson in His Journals* 5.

38. Emerson, "Intellect," in *The Complete Essays* 292.

39. Ibid., 292–293.

40. Willis, "Inaugural Address," liii.

41. Quoted in *The Christian Examiner* 66 (March 1859): 223.

42. Peabody, *Sermon Preached at Portland, Maine* 7.

43. Willis, addendum to Peabody, *Sermon* 25.

44. Nichols, *A Catechism of Natural Theology.*

45. Nichols, *An Oration* 18.

46. Ibid., 13.

47. Ibid., 8.

48. Anonymous, *The Christian Examiner* 66 (March 1859): 229.

49. Byerly, "Reminiscences," 6.

50. Charles Sanders Peirce, Notes on Benjamin Peirce, unpublished two-page sheet found in Charles Sanders Peirce's copy of Jordan's *Traité des Substitutions et des Equations Algébriques,* 28 June 1910, Charles Sanders Peirce Papers, Houghton Library, Harvard University, Cambridge.

51. Peirce, *Ideality in the Physical Sciences* 28–29.

52. Byerly, "Reminiscences," 6.

53. Sven R. Peterson, "Benjamin Peirce: Mathematician Philosopher," in Archibald, *Benjamin Peirce, 1809–1880* 98–99.

54. Peirce, *Address to the American Association for the Advancement of Science* 2–3.

55. Ibid., 3–4.

56. Ibid., 3.

57. Ibid., 4.

58. Ibid., 5.

59. Ibid., 14.

60. Ibid., 6.

61. Ibid., 8–9.

62. Ibid., 6.

63. Peirce, *Linear Associative Algebra* preface (unnumbered page).

64. The formula, using an idiosyncratic notation, was $\Sigma[m,(D,v,Ss\text{-}\Sigma'F,Sf)]$. Benjamin Peirce to Ichabod Nichols, 9 September 1858, 1861 copy with Ichabod Nichols's response, Benjamin Peirce Papers, Houghton Library, Harvard University, Cambridge.

65. Ibid.

66. Quoted in Emerson, *The Early Years of the Saturday Club* 105.

67. Land, *Thomas Hill* 51.

68. See Hill's letter to his sister Henrietta, quoted in Land, *Thomas Hill* 53–54.

69. Emerson, *The Early Years of the Saturday Club* 99.

70. Land, *Thomas Hill* 4–5. Additional biographical material can be found in Allen, *Sequel to 'Our Liberal Movement'*, and parts of Hill's own article "Books That Have Helped Me," 388–396.

71. Hill, "Books That Have Helped Me," 390.

72. Quoted in Land, *Thomas Hill* 9–10.

73. Hill, "Books That Have Helped Me," 391.

74. Ibid., 392, 394.

75. Agassiz, *Contributions to the Natural History of the United States* 1:135.

76. Hill, "Books That Have Helped Me," 392. See also Land, *Thomas Hill* 197, 213.

77. Quotation on Hill's first reading of Emerson's *Nature* is from Land, *Thomas Hill* 28.

78. Hill, "Books That Have Helped Me," 393.

79. Ibid., 394.

80. Hill, *Philosophy Higher than Science* 6.

81. The survey of Hamilton's thought commonly available to Americans was the compilation *Philosophy of William Hamilton*.

82. Quoted in Hill, *The Natural Sources of Theology* 4.

83. Ibid., 13.

84. Ibid.

85. Ibid., 14.

86. Hill, *Geometry and Faith*, 3rd ed., 25.

87. For example, from John Dee: "No man, therefore, can doubt, but toward the attaining of knowledge incomparable, and Heavenly wisdom, mathematical speculations, both of Numbers and Magnitudes, are means, aids and guides; ready, certain and necessary." Hill, *Geometry and Faith* frontispiece.

88. Hill, *Geometry and Faith*, 3rd ed., 8.

89. Ibid., 62.

90. Ibid., 108.

91. Ibid., 21.

92. On this shift in university education, see Olsen and Voss, *The Organization of Knowledge in Modern America 1860–1920* and Reingold, *Science in Nineteenth-Century America*.

93. Hill, *Liberal Education* 20.

94. Hill, *The True Order of Studies* 25–26.

95. Hill, *First Lessons in Geometry* 8.

96. Ibid., 135.

97. Ibid., 135–136.

98. Hill, *Religion in Public Instruction* 6.

99. Ibid., 8.

100. Hill, *Annual Report of the President and Treasurer of Harvard College* 1864, 7.

101. Ibid., 8.

102. Hill, *Annual Report of the President and Treasurer of Harvard College* 1865, 9.

103. Ibid., 11.

104. Paper read before the Harvard Corporation, quoted in Land, *Thomas Hill* 145.

105. See James, "Engineering an Environment for Change," 55–75, esp. 65–66 and 70–72. James also fails to mention Peirce's idealist vision of mathematics.

106. Peirce, "The Intellectual Organization of Harvard," 77.

107. Ibid.

108. Hoar, "Harvard College Fifty-Eight Years Ago," 64.

109. Quoted in Cajori, *The Teaching and History of Mathematics in the United States* 121.

110. Quoted in Archibald, *Benjamin Peirce, 1809–1889* 18. Outside of Harvard Peirce, in fact, had a few female students, set up a class for women in geometry, and was a constant advocate of coeducation. See H. T. W., "Benjamin Peirce," *Woman's Journal,* 23 October 1880.

111. Charles Eliot, "Reminiscences of Peirce," in Archibald, *Benjamin Peirce, 1809–1880* 2–3.

112. "The Late Professor Peirce," *The Nation,* 14 October 1880, 268. Reprinted in *The Harvard Register* 3 (1881): 56.

113. Emerson, *The Early Years of the Saturday Club* 102.

114. *Scientific Monthly* 3 (1916): 616.

115. Wait, "Advanced Instruction in American Colleges," 127.

116. "The Late Professor Peirce," 268. This comment probably stems from an incident at a National Academy of Sciences meeting in the early 1870s. According to Thomas Hill, Peirce had just finished reading from his recently published work *Linear Associative Algebra* when Louis Agassiz stood up in awe and commented, "I must confess that I have not understood one word of this communication, but I have heretofore had such ample reasons for believing in the speaker's clearness and soundness of thought, that I accept what he has now said as undoubtedly true." Hill, *The Natural Sources of Theology* 102.

117. Emerson, *The Early Years of the Saturday Club* 97; Byerly, "Reminiscences," 5.

118. Peirce, *Linear Associative Algebra* 1.

119. Peirce, "Address to the American Association for the Advancement of Science," 11. He did not name these opponents.

120. Ibid., 12.

121. Quoted in Emerson, *The Early Years of the Saturday Club,* 105.

122. See Roberts, *Darwinism and the Divine in America* esp. 117–145, for American responses to evolution that were similar to Peirce's.

123. Peirce, "The Conflict Between Science and Religion," 656.

124. This squaring of the Bible with evolution became common among sci-

entifically literate American intellectuals. See Roberts, *Darwinism and the Divine in America* 147–148.

125. Peirce, *Ideality in the Physical Sciences* 48.

126. Ibid., 56–57. Emphasis in the original.

127. Ibid., 194. Emphasis in the original.

128. William E. Copeland, "Spiritualism of Science," *The Monthly Religious Magazine* 49 (1873): 40.

129. Peirce, *Ideality in the Physical Sciences* 32.

130. Hill, *God Known by His Works* 4.

131. See Cashdollar, *The Transformation of Theology* 11.

132. Thomas Hill, Review of *Logic*, by John Stuart Mill, 366–368.

133. Cashdollar, *The Transformation of Theology* 110.

134. Hill, *Philosophy Higher Than Science* 6–7.

135. Hill, *God Known by His Works* 3.

136. Hill, *The Natural Sources of Theology* 2.

137. Hill, *Religion in Public Instruction* 15.

138. Hill, *God Known by His Works* 10.

139. Ibid., 11.

140. Running from January 1874 to April 1875.

141. Hill, *God Known by His Works* 11.

142. Hill, *The Natural Sources of Theology* 72.

143. Ibid., 66. Joseph Henry Allen recalled another example of this coincidence between pure mathematics and the construction of nature: "Calling upon [Hill] one day at the President's office, I found him engaged for some few minutes, and, to while away the time, he asked me to contemplate the following formula, $p = ar$, and see what I could make of it . . . He then explained the formula, showing how, by assigning different arbitrary values to a, a wonderful variety of curves could be developed, some of them extremely intricate and beautiful . . . And he told me how Benjamin Peirce, that prince of mathematicians, in whom imagination and reverence kept pace with all the movements of his thought, found him once engaged in these constructions, and, being fascinated by the theory, brought in [Louis] Agassiz to see; and Agassiz, his eye being caught by one of the forms, exclaimed, 'Why, that is the very shape taken at one stage of its growth in the nerve-cord of a crab!'" Allen, *Sequel to 'Our Liberal Movement'* 127.

144. Hill, *The Natural Sources of Theology* 66–67.

145. Ibid., 67.

146. Hill, *Religion in Public Instruction* 8.

147. Hill, *Philosophy Higher Than Science* 7.

148. Ibid.

149. Ibid., 10–11.

150. Ibid., 11.

151. Ibid., 10.

152. Ibid., 4.

153. Ibid., 12.

154. For Norris's use of the term see *An Essay Towards the Theory of the Ideal or Intelligible World* 1:166, 177.

155. Peirce, *Ideality in the Physical Sciences* 15–16.

156. Ibid., 17.

157. Ibid., 165.

158. Ibid., 167.

159. Ibid., 194.

160. Ibid., 26. Emphasis mine.

161. Ibid., 19.

162. Ibid., 188–189.

163. Ibid., 36.

164. Ibid., 37.

165. Coolidge, "The Story of Mathematics at Harvard," 374.

166. Peirce, "On the exact fractions which occur in phyllotaxis," quoted in Archibald, *Benjamin Peirce, 1809–1880* 12.

167. See, e.g., "Benjamin Peirce," *The Boston Daily Advertiser,* 7 October 1880, 1.

CHAPTER THREE George Boole and the Genesis of Symbolic Logic

1. See M. E. Boole, "Home-side," 28; Mary Everest Boole, "Indian Thought and Western Science in the Nineteenth Century," in *Collected Works* 3:951.

2. M. E. Boole, "Home-side," 40.

3. Bertrand Russell, Untitled tribute to George Boole, typescript, n.d., Royal Irish Academy, Dublin.

4. George Boole to Charles Taylor, 27 April 1840, typescript, BP/1/226(3), Boole Papers, University College Library, University College, Cork. The University College, Cork, collection is hereafter referred to as Boole Papers, UCC.

5. At one point Boole was so enamored with Kant's thought that he recommended one of the German philosopher's works at the Lincoln Mechanics Institute. This suggestion stunned his friend T. Bainbridge, who reproachfully declared, "Immanuel Kant! O George, George, I did not expect that from you!" William Brooke to Maryann Boole, n.d., BP/1/249, Boole Papers, UCC.

6. For a complete biography of Boole, see MacHale, *George Boole: His Life and Work.*

7. Anonymous letter to Maryann Boole, 18 March 1853, BP/1/267, Boole Papers, UCC.

8. J. Dyson to Maryann Boole, n.d. though probably a recollection from 1865–1866 following George Boole's death, BP/1/256, Boole Papers, UCC.

9. George Boole to Charles Kirk, 21 March 1848, BP/1/222(7), Boole Papers, UCC.

10. George Boole to Rev. E. Larken, 7 July 1848, BP/1/223(14), Boole Papers, UCC.

11. George Boole, "Sonnet XIII," 3 July 1849, typescript, Ms 12.K.45(vi), Royal Irish Academy, Dublin.

12. George Boole, "Sonnet XII, Carisbrooke, To I.P.," July 1849, typescript, Ms 12.K.45(vi), Royal Irish Academy, Dublin.

13. Newspaper clipping (probably from *The Stamford Record*), n.d., Ms UP 9663, Local History Collection, Lincolnshire Central Library, Lincoln.

14. George Boole to Maryann Boole, 25 October 1849 and 30 March 1850, BP/1/6 and BP/1/28, Boole Papers, UCC.

15. Boole, "Sonnet XV," 2 November 1849, typescript, Ms 12.K.45(vi), Royal Irish Academy, Dublin.

16. George Boole to Mary Boole, 20–25 March 1850, BP/1/150, Boole Papers, UCC.

17. George Boole to Augustus De Morgan, 8 November 1849, in Smith, *The Boole-De Morgan Correspondence* 24.

18. George Boole to Maryann Boole, 29 May 1850, BP/1/34, Boole Papers, UCC.

19. Newspaper clipping (probably from *The Stamford Record*), Ms UP 9663, Local History Collection, Lincolnshire Central Library, Lincoln.

20. On some of the background to the establishment of both the Queen's Colleges and Catholic University, see Garland, "Newman in His Own Day," 265–281.

21. George Boole to Maryann Boole, 26 February 1850, BP/1/24, Boole Papers, UCC.

22. Quoted in Ralls, "The Papal Aggression of 1850," 127.

23. George Boole to Maryann Boole, 15 October 1850, BP/1/41, Boole Papers, UCC.

24. de Vericour, *An Historical Analysis of Christian Civilisation* 18.

25. George Boole to Maryann Boole, 25 August 1850, BP/1/40, Boole Papers, UCC.

26. Hales, *The Catholic Church in the Modern World* 108.

27. Ralls, "The Papal Aggression of 1850," 115–134; Machin, *Politics and the Churches in Great Britain 1832–1868* 210–8; Chadwick, *The Victorian Church* 1:271–309.

28. George Boole to Mary Boole, 23 October 1849, BP/1/140, Boole Papers, UCC.

29. George Boole to Mary Boole, 1 November 1849, BP/1/142, Boole Papers, UCC.

30. George Boole to Maryann Boole, 3 November 1850, BP/1/44, Boole Papers, UCC.

31. George Boole to Maryann Boole, 27 February 1852, BP/1/75, Boole Papers, UCC.

32. George Boole to Augustus De Morgan, 17 October 1850, in Smith, *The Boole-De Morgan Correspondence* 38.

33. George Boole to Maryann Boole, 18 November 1850, BP/1/46, Boole Papers, UCC.

34. George Boole to Maryann Boole, 2 December 1850, BP/1/49, Boole Papers, UCC.

35. George Boole to Maryann Boole, 3 November 1850, BP/1/44, Boole Papers, UCC.

36. George Boole to Maryann Boole, 13 January 1851, BP/1/55, Boole Papers, UCC.

37. George Boole to Maryann Boole, 18 November 1850, BP/1/46, Boole Papers, UCC.

38. George Boole to Maryann Boole, 6 March 1853, BP/1/94, Boole Papers, UCC.

39. George Boole to William Brooke, 18 June 1855, typescript, BP/1/161(a), Boole Papers, UCC.

40. Ibid.

41. George Boole to Maryann Boole, 25 May 1857, BP/1/125, Boole Papers, UCC.

42. Ibid.

43. George Boole to Maryann Boole, 18 November 1850, BP/1/46, Boole Papers, UCC.

44. M. E. Boole, "Home-Side," 2.

45. Ibid.

46. George Boole to William Brooke, n.d., BP/1/160, Boole Papers, UCC.

47. George Boole to A. T. Taylor, 26 January 1860, BP/1/232, Boole Papers, UCC.

48. Newman, *Phases of Faith* 233–234.

49. George Boole to E. Larken, 31 May 1847, BP/1/223(8), Boole Papers, UCC.

50. Unpublished recollection by T. A. A. Broadbent, n.d. though probably following George Boole's death in 1864, BP/1/318, Boole Papers, UCC. See also M. E. Boole, "Home-side," 14.

51. George Boole to Joseph Hill, 17 February 1844, BP/1/221(13), Boole Papers, UCC.

52. M. E. Boole, "Home-side," 15.

53. Charles Clarke to Maryann Boole, 17 December 1865, BP/1/254, Boole Papers, UCC.

54. Ibid.

55. M. E. Boole, "Home-side," 38.

56. George Boole, "Sonnet to the Number Three," 16 May 1846, typescript, BP/1/305(1), Boole Papers, UCC.

57. M. E. Boole, "Home-side," 41–42.

58. George Boole to M. C. Taylor, April 1840, BP/1/226, Boole Papers, UCC.

59. Charles Clarke to Maryann Boole, 17 December 1865, BP/1/254, Boole Papers, UCC.

60. George Boole, "Are the Planets Inhabited," unpublished lecture, BP/1/272, Boole Papers, UCC.

61. Laurence Elvin, unpublished biography of George Boole, Ms U.P. 3721, p. 1, Local History Collection, Lincolnshire Central Library, Lincoln.

62. Charles Clarke to Maryann Boole, 17 December 1865, BP/1/254, Boole Papers, UCC.

63. George Boole, "Sonnet III," June 1849, typescript, Ms 12.K.45(vi), Royal Irish Academy, Dublin.

64. See also the unpublished poems "Sonnet V" (June 1849), "Sonnet X" (5 July 1849), and "Sonnet" (1841), Ms 12.K.45(vi), Royal Irish Academy, Dublin.

65. George Boole, "The Right Use of Leisure," an address delivered to the members of the Lincoln Early Closing Association, 9 February 1847, BP/1/279, pp. 4–5, Boole Papers, UCC.

66. George Boole to Charles Taylor, 27 April 1840, typescript, BP/1/226(3), Boole Papers, UCC.

67. George Boole to Augustus De Morgan, 17 October 1850, in Smith, *The Boole-De Morgan Correspondence* 38.

68. Boole, "Sonnet to the Number Three."

69. George Boole, "2 Peter 3.13," n.d., Ms 12.K.45(vii), Royal Irish Academy, Dublin.

70. George Boole, "Stanzas," n.d., BP/1/305(1), Boole Papers, UCC.

71. M. E. Boole, "Home-side," 42.

72. Boole, "The Right Use of Leisure," 19–20.

73. M. E. Boole, "Home-side," 46. Boole used the verse as a title for an unpublished poem dated 1 December 1849, Ms 12.K.45(vi), Royal Irish Academy, Dublin.

74. M. E. Boole, "Home-side," 26.

75. Ibid., 7, 29.

76. Boole, "The Right Use of Leisure," 19–20.

77. George Boole to Maryann Boole, n.d., BP/1/134, Boole Papers, UCC.

78. M. E. Boole, "Home-side," 38.

79. On his Catholic Church attendance, see George Boole to Maryann Boole, 12–13 December 1849, BP/1/16, Boole Papers, UCC; On his attendance at independent churches see George Boole to Mary Boole, 21 November 1849, BP/1/143, Boole Papers, UCC.

80. George Boole to Maryann Boole, 30 January 1850, BP/1/20, Boole Papers, UCC.

81. George Boole to Maryann Boole, 9 November 1850, BP/1/45, Boole Papers, UCC.

82. George Boole to Joseph Hill, 30 November 1840, BP/1/221(25), Boole Papers, UCC.

83. George Boole to M.C. Taylor, 28 February 1845, BP/1/230, Boole Papers, UCC.

84. Ibid.

85. M. E. Boole, "Home-side," 3.

86. Ibid.

87. Ibid., 39.

88. George Boole to Charles Kirk, 24 January 1849, BP/1/222(10), Boole Papers, UCC.

89. George Boole, "Lines Written in the Autumn of 1846," typescript, Ms 12.K.45 (vii), Royal Irish Academy, Dublin.

90. George Boole to Joseph Hill, 11 December 1850, BP/1/221(23), Boole Papers, UCC.

91. George Boole to E. Larken, 29 April 1847, BP/1/223(7), Boole Papers, UCC.

92. George Boole to E. Larken, 23 January 1847, BP/1/223(6), Boole Papers, UCC.

93. M. E. Boole, "Home-side," 16–17, 31.

94. Ibid., 7.

95. Ibid., 42. Emphasis in the original.

96. Ibid., 47.

97. See Jeff Guy, *The Heretic: A Study of the Life of John William Colenso, 1814–1883,* and A. L. Rowse, *The Controversial Colensos.*

98. M. E. Boole, "Home-side," 45.

99. First name Bernard, last name perhaps Arbarbanell. Writing to another childhood mentor and friend, William Brooke, Boole mentions "our friend of yore Bernard, replete with all the Talmud's lore" (George Boole to William Brooke, n.d., BP/1/300, Boole Papers, UCC). Mary Everest Boole mentions a Dr. Arbarbanell of "Jewish descent" in "Home-side," 7.

100. George Boole to Joseph Hill, 11 February 1837 and 30 May 1837, BP/1/221(5) and BP/1/221(6), Boole Papers, UCC; Joseph Hill to Maryann Boole, 20 March 1866 and 24 May 1866, BP/1/259 and BP/1/260, Boole Papers, UCC.

101. George Boole to Joseph Hill, 30 May 1837, BP/1/221(6), Boole Papers, UCC.

102. George Boole to Augustus De Morgan, 4 November 1861, Ms Add. 97, De Morgan Papers, Special Collections, University College Library, University College, London.

103. M. E. Boole, "Home-side," 41.

104. Ibid., 24.

105. Charles Clarke to Maryann Boole, 17 December 1865, BP/1/254, Boole Papers, UCC.

106. Boole, "Lines Written in the Autumn of 1846."

107. M. E. Boole, "Home-side," 7.

108. George Boole to Maryann Boole, 14 April 1854, BP/1/119, Boole Papers, UCC.

109. George Boole to Augustus De Morgan, 8 October 1852, in Smith, *The Boole-De Morgan Correspondence* 62.

110. Mary Bendorf to Mary Boole, n.d., BP/1/338, Boole Papers, UCC.

111. Boole, *The Laws of Thought* 31. Emphasis in the original.

112. Boole's use of the symbol "1" for the universe of discourse seems influenced by Jewish monotheism.

113. Aristotle, *Metaphysics* 161–163 (1005b).

114. Boole, *The Laws of Thought* 84.

115. Ibid., 215–216.

116. Ibid., 187.

117. Ibid., 188.

118. Ibid., 189–191. That is, Clarke's proof is a tautology.

119. Ibid., 191.

120. Ibid., 185–186.

121. Ibid., 20–21.

122. Ibid., 424.

123. On Newman, see George Boole, Undated Ms, C1.6, Boole Papers, Royal Society of London, London; On Plato, see George Boole, Undated Ms, C.36–37, Boole Papers, Royal Society of London, London. On theodicy, see George Boole, Undated Ms, C.35, Boole Papers, Royal Society of London, London.

124. George Boole, Undated Ms, B.124, Boole Papers, Royal Society of London, London.

125. George Boole, Undated Ms, B.125, Boole Papers, Royal Society of London, London.

126. Boole, *The Laws of Thought* 217–218.

CHAPTER FOUR Augustus De Morgan and the Logic of Relations

1. Augustus De Morgan, Notebook, Ms Add. 69, De Morgan Papers, Special Collections, University College Library, University College, London.

2. Ibid.

3. De Morgan, "On the Syllogism: IV and on the Logic of Relations," 208–246.

4. De Morgan used the anagram many times, including in his letter to John Herschel, 3 September 1866, Ms HS.6.387, Herschel Papers, Royal Society of London, London.

5. For a technical discussion of De Morgan's advances in logic and mathematics, see Merrill, *Augustus De Morgan and the Logic of Relations;* for the broadest biography, see Sophia Elizabeth De Morgan, *Memoir of Augustus De Morgan,* hereafter referred to as S. De Morgan, *Memoir;* for shorter accounts of De Morgan's life and work, see Stirling, *William De Morgan and His Wife;* Howson, *A History of Mathematics Education in Modern England* 75–96; Peter Heath's introduction to his edited edition of De Morgan's work *On the Syllogism and Other Logical Writings;* and Crowther, *Scientific Types* 189–223.

6. S. De Morgan, *Memoir* 1.

7. De Morgan to William Rowan Hamilton, 2 February 1852, Ms 1493/541, Hamilton Papers, Special Collections, Trinity College Library, Trinity College, Dublin.

8. S. De Morgan, *Memoir* 86.

9. Ibid., 10.

10. Ibid., 14. Emphasis in the original. When possible, letters published in Sophia De Morgan, *Memoir of Augustus De Morgan*, were checked with the originals for accuracy. The only apparent difference is that Sophia De Morgan italicized words or phrases that were underlined in the handwritten originals.

11. Ibid., 15.

12. Ibid., 115–6.

13. See, e.g., Augustus De Morgan to Alexander Campbell Fraser, 24 December 1863 and 16 January 1864, Ms Deposit 208, Box 17, Special Collections, National Library of Scotland, Edinburgh.

14. Augustus De Morgan, Autobiographical sketch written in the third person, Ms 28509, f. 421, Special Collections, British Library, London. De Morgan lamented his lack of a master's degree for the rest of his life. "Knew you not that I am a heretic who is B.A. of thirty-one years' standing by reason of subscriptions being unsubscribable?" he regretfully wrote to John Herschel in 1858 (Augustus De Morgan to John Herschel, 15 November 1858, in S. De Morgan, *Memoir* 297). De Morgan had written to William Rowan Hamilton six years earlier, "When I came to think about my M.A. degree, I found I should be required to declare that all who dealt wrongly with substance and person would perish everlastingly; and so I continue B.A." (Augustus De Morgan to William Rowan Hamilton, 16 May 1852, in Graves, *William Rowan Hamilton* 3:364. When possible, De Morgan's letters to Hamilton were checked with the originals for accuracy.).

15. De Morgan to William Rowan Hamilton, 27 July 1852, in Graves, *William Rowan Hamilton* 3:395.

16. Originally called the University of London; upon the creation of an umbrella administration, which included King's College, the original University of London became University College, London. See Willson, *Our Minerva: The Men and Politics of the University of London, 1836–1858*; Desmond, *The Politics of Evolution* 25–41.

17. Harte and North, *The World of UCL, 1828–1990* 10–12.

18. Ibid., 19.

19. Quoted in ibid., 31.

20. *John Bull* 23 January 1826, 29.

21. S. De Morgan, *Memoir* 183.

22. Quoted in Desmond, *The Politics of Evolution* 25.

23. Augustus De Morgan, Letter to the Council, 1838, 4438c, College Correspondence, Special Collections, University College Library, University College, London.

24. Augustus De Morgan to George Boole, 16 October 1861 and 21 November 1861, in Smith, *The Boole-De Morgan Correspondence* 87, 93.

25. S. De Morgan, *Memoir* 102.

26. Harte and North, *The World of UCL* 39.

27. Pattison, *Granville Sharp Pattison* 153ff.

28. Augustus De Morgan, Letter to the Council, March 1830, 3350d, College Correspondence, Special Collections, University College Library, University College, London.

29. Augustus De Morgan, Letter to the Council, 3350e–h, College Correspondence, Special Collections, University College Library, University College, London.

30. Augustus De Morgan, Letter to the Council, 2470, College Correspondence, Special Collections, University College Library, University College, London.

31. Augustus De Morgan, Letter to the Council, 2108, College Correspondence, Special Collections, University College Library, University College, London.

32. Desmond, *The Politics of Evolution* 26–27; Berman, "'Hegemony' and the Amateur Tradition in British Science," 30–50.

33. See Percival, *The Society for the Diffusion of Useful Knowledge;* Grobel, "The Society for the Diffusion of Useful Knowledge 1826–1846 and its Relation to Adult Education in the First Half of the XIXth Century."

34. See Morrell and Thackray, *Gentlemen of Science* 2–21, 224–296.

35. S. De Morgan, *Memoir* 52.

36. Ibid., 117.

37. Ibid., 143.

38. See Coleridge, *Confessions of an Inquiring Spirit.* On the influence of Strauss in England, see Chadwick, *The Victorian Church* 1:528–544.

39. S. De Morgan, *Memoir* 139.

40. Ibid.

41. Ibid., 140. Emphasis in the original.

42. Ibid., 141.

43. Ibid., 142.

44. Augustus De Morgan to W. Heald, 18 August 1851, in S. De Morgan, *Memoir* 216.

45. Ibid.

46. Augustus De Morgan to William Rowan Hamilton, 10 May 1852, in Graves, *William Rowan Hamilton* 3:362.

47. Knight, *University Rebel* 304, 306–307.

48. See Smith, *The Boole-De Morgan Correspondence.*

49. Augustus De Morgan to George Boole, 14 August 1849, in Smith, *The Boole-De Morgan Correspondence* 35.

50. Augustus De Morgan to William Rowan Hamilton, 26 September 1849, in Graves, *William Rowan Hamilton* 3:277.

51. Augustus De Morgan, Letter to the UCL Council, 5 November 1853, in S. De Morgan, *Memoir* 187.

52. S. De Morgan, *Memoir* 188–189.

53. Years after his death, William Hamilton's disciples were still pestering Augustus De Morgan about his "theft" of the quantification of the predicate and his insistence that philosophers would have to cede the field of logic to mathematicians.

See, e.g., De Morgan's complaints to Alexander Campbell Fraser throughout 1862: 16 March, 23 May, 18 October, and 26 December, Deposit 208, Box 17, Special Collections, National Library of Scotland, Edinburgh.

54. Augustus De Morgan, Note between pp. 328 and 329 of William Hamilton's returned copy of De Morgan's *Formal Logic*, n.d., Ms 776/1, De Morgan Papers, Special Collections, University of London Library, London.

55. De Morgan kept a scrapbook of what he called "The Gorilla War," Ms 775/366, De Morgan Papers, Special Collections, University of London Library, London.

56. Whewell, *Thoughts on the Study of Mathematics as a Part of a Liberal Education*. The work was one of the first treatises calling for an increase in mathematical education at the universities in the wake of the Cambridge Analytical Society.

57. Hamilton, *Discussions on Philosophy and Literature* 268. Originally published as "On the Study of Mathematics," *Edinburgh Review* 126 (January 1836): 409–455. Emphasis in the original.

58. Ibid., 268–272.

59. Ibid., 275. Emphasis in the original.

60. Ibid., 281–282.

61. Augustus De Morgan to William Rowan Hamilton, 12 February 1853, in Graves, *William Rowan Hamilton* 3:447.

62. Augustus De Morgan to John Herschel, 22 November 1842, in S. De Morgan, *Memoir* 147. Emphasis in the original.

63. Augustus De Morgan, Note on p. 27 of William Hamilton's returned copy of De Morgan's *Formal Logic*, 2 May 1850, Ms 776/1, De Morgan Papers, Special Collections, University of London Library, London.

64. De Morgan believed that he had found the perfect summary of metaphysics in four lines from William Wordsworth's *Pastoral Hymns:* "Here we go up, up, up,/ and there we go down, down, down,/Here we go backwards and forwards/And there we go round round round." Augustus De Morgan, "Preface" to an unreferenced work (possibly *From Matter to Spirit* by Sophia De Morgan), n.d., Ms Add. 163, De Morgan Papers, Special Collections, University College Library, University College, London.

65. Jotted on an interleaved copy of his *Formal Logic* in 1847, S. De Morgan, *Memoir* 164–165. As De Morgan had snidely written to William Whewell a year earlier, a highly effective way to convince someone that Berkeley was wrong was to physically assault them (Augustus De Morgan to William Whewell, 21 October 1846, Add. Ms A.202106, Whewell Papers, Trinity College Library, Trinity College, Cambridge).

66. Augustus De Morgan to John Herschel, 15 October 1858, Ms HS.6.317, Herschel Papers, Royal Society of London, London.

67. As De Morgan later wrote, "I have looked for ten years at various ontological writings about 'the unconditioned' and various religious works about the Almighty, and I think I see a very great tendency to confuse omnipresent personality with infinite extent. At least there is a want of power to put the distinction into language."

Augustus De Morgan to William Whewell, 1 April 1863, Add. Ms A.202147, Whewell Papers, Trinity College Library, Trinity College, Cambridge.

68. A joke; William Rowan Hamilton and William Hamilton were, of course, unrelated. De Morgan enjoyed ribbing his friend about how he shared the same name as his nemesis.

69. Augustus De Morgan to William Rowan Hamilton, 16–17 July 1852, in Graves, *William Rowan Hamilton* 3:386.

70. Augustus De Morgan to Alexander Campbell Fraser, 23 November 1862, Deposit 208, Box 17, Special Collections, National Library of Scotland, Edinburgh.

71. De Morgan, *Formal Logic* 46–47.

72. Ibid., 182–183.

73. Ibid., 237–286.

74. Ibid., 262. Emphasis in the original.

75. Ibid.

76. For example, in politics, where "an argument in favour of checking the power of the Crown is called Jacobinism; of an increase of that power, absolutism." Ibid., 263.

77. Ibid., 241.

78. Locke, *An Essay Concerning Human Understanding* 475–476; for Locke's entire discussion of the misuse and abuse of language, see 474–524. On Locke's general concern with language and ethics, see Wolterstorff, *John Locke and the Ethics of Belief.*

79. Augustus De Morgan, preface to Sophia De Morgan, *From Matter to Spirit* xxix–xxx.

80. Augustus De Morgan, Introductory lecture in mathematics, 1828, p. 16, Ms Add.3, De Morgan Papers, Special Collections, University College Library, University College, London.

81. For a more technical discussion of this system, see Merrill, *Augustus De Morgan and the Logic of Relations* esp. 63ff.

82. De Morgan to William Rowan Hamilton, 1 September 1852, in Graves, *William Rowan Hamilton* 3:410. The quip was reiterated in A. De Morgan, preface to S. De Morgan, *From Matter to Spirit* vii.

83. Augustus De Morgan to William Rowan Hamilton, 23 August 1852, in Graves, *William Rowan Hamilton* 3:404. Emphasis in the original.

84. Augustus De Morgan to William Whewell, 3 March 1860, Add. Ms A.202143, Whewell Papers, Trinity College Library, Trinity College, Cambridge; Augustus De Morgan to William Rowan Hamilton, 16 May 1852, in Graves, *William Rowan Hamilton* 3:364.

85. Augustus De Morgan, Note opposite p. 68 of William Hamilton's returned copy of De Morgan's *Formal Logic,* n.d., Ms 776/1, De Morgan Papers, Special Collections, University of London Library, London.

86. See, e.g., letter from "Orthodox" to Augustus De Morgan, n.d., Ms 775/370/27, De Morgan Papers, Special Collections, University of London Library, London.

87. S. De Morgan, *Memoir* 364.

88. De Morgan, *Address to the London Mathematical Society* 4. Unfortunately for De Morgan, the young mathematicians who took his advice ultimately abandoned his awkward system in favor of Boole's more effective notation.

89. Ibid., 8.

90. Oppenheim, *The Other World* 37, 335–336.

91. De Morgan to William Rowan Hamilton, 12 July 1847, Ms 1493/402, Hamilton Papers, Special Collections, Trinity College Library, Trinity College, Dublin.

92. Augustus De Morgan to William Heald, 1849, in S. De Morgan, *Memoir* 206–208.

93. Augustus De Morgan to William Rowan Hamilton, 16 April 1852, in Graves, *William Rowan Hamilton* 3:352.

94. See Oppenheim, *The Other World* 11; Barrow, *Independent Spirits* 1–18; Gauld, *The Founders of Psychical Research* 66–70.

95. Gauld, *The Founders of Psychical Research* 67.

96. A. De Morgan, preface to S. De Morgan, *From Matter to Spirit* xli–xlv. The De Morgans' testimony remained an important touchstone for spiritualists for more than a generation. See, e.g., Doyle, *The History of Spiritualism* 2:298–302.

97. Augustus De Morgan to William Heald, July 1853, in S. De Morgan, *Memoir* 221–222; A. De Morgan, preface to S. De Morgan, *From Matter to Spirit* xliii.

98. De Morgan to William Rowan Hamilton, 9 May 1853, in Graves, *William Rowan Hamilton* 3:450–451.

99. Although it was published anonymously under the pseudonyms A.B. and C.D., the De Morgans confirmed to friends what many in London intellectual circles had already suspected. "In a few days will be published a book by C.D. with a preface by A. B. People say that C.D. is Mrs. De Morgan and that A.B. is myself—and I do not intend to deny either assertion," Augustus De Morgan wrote to Henry Brougham on 23 October 1863 (Ms A.26.422, Brougham Papers, Special Collections, University College Library, University College, London). See also Augustus De Morgan to George Boole, 5 July 1863, in Smith, *The Boole-De Morgan Correspondence* 112.

100. A. De Morgan, preface to S. De Morgan, *From Matter to Spirit* v.

101. See Oppenheim, *The Other World* 59–62; Gauld on "reluctant doubters," in *The Founders of Psychical Research* 32–65.

102. Augustus De Morgan to Henry Brougham, 17 June 1855, Ms 4712, Brougham Papers, Special Collections, University College Library, University College, London.

103. See, e.g., Augustus De Morgan to Samuel Maitland, 1853, Ms 913, box 2, packet 3, De Morgan Papers, Special Collections, University of London Library, London. Also see S. De Morgan, *Memoir* 124–125.

104. "I am much obliged to you for your little work, which is well adopted to excite inquiry," De Morgan wrote to Wallace when the latter finally published a monograph on mediums and clairvoyants in 1866, "Your book will set many rational persons suspecting they ought to inquire [about psychic forces]." Augustus De Morgan

to Alfred Russel Wallace, 1866, ff. 8–9, Ms 46439, Special Collections, British Library, London.

105. Augustus De Morgan to Alexander Campbell Fraser, 21 June 1855, Ms Deposit 208, Box 17, Special Collections, National Library of Scotland, Edinburgh.

106. See, e.g., Gauld, *The Founders of Psychical Research* 75.

107. See Berman, *Social Change and Scientific Organization* 158–159, and Cantor, *Michael Faraday* 294.

108. Quoted in Thomas, *Michael Faraday and the Royal Institution* 127.

109. See Jones, *The Life and Letters of Faraday* 2:307–308, 318, 468–470; Faraday, *The Selected Correspondence of Michael Faraday* 2:874–875.

110. Faraday, "On Table-Turning" and "Experimental Investigation of Table-Turning," in *Experimental Researches* 463–491.

111. Cantor, *Michael Faraday* 200.

112. Augustus De Morgan to William Rowan Hamilton, 1 October 1852, in Graves, *William Rowan Hamilton* 3:418.

113. Cantor, *Michael Faraday* 216.

114. These criticisms culminated in a strongly antimathematical report to a Parliamentary education committee. See Michael Faraday, "Report of H.M. Commissioners appointed to inquire into the revenues and management of certain colleges and schools and the studies pursued and instruction given therein; with appendix and evidence," *Sessional Papers* London: House of Commons, 1864, 4:377.

115. Tyndall, *Faraday as a Discoverer* 63–64.

116. Cantor, *Michael Faraday* 219–220.

117. Faraday, "Observations on Mental Education," in *Experimental Researches* 482–491.

118. Quoted in S. De Morgan, *Memoir,* 192.

119. Ibid.

120. Augustus De Morgan, Introductory lecture in mathematics, 1828, p. 19, Ms Add.3, De Morgan Papers, Special Collections, University College Library, University College, London.

121. A. De Morgan, preface to S. De Morgan, *From Matter to Spirit* xix–xx.

122. Ibid., xviii.

123. De Morgan, *A Budget of Paradoxes* 378.

124. A. De Morgan, preface to S. De Morgan, *From Matter to Spirit* vi.

125. Ibid., vi.

126. Ibid.

127. Augustus De Morgan to William Rowan Hamilton, 22 January 1852, Ms 1493/529, Hamilton Papers, Special Collections, Trinity College Library, Trinity College, Dublin.

128. Augustus De Morgan to William Rowan Hamilton, 12 February 1853, in Graves, *William Rowan Hamilton* 3:447.

129. Augustus De Morgan to William Rowan Hamilton, 10 December 1853, in Graves, *William Rowan Hamilton* 3:467.

130. Augustus De Morgan to William Rowan Hamilton, 3 March 1856, in Graves, *William Rowan Hamilton* 3:507. Emphasis in the original.

131. Augustus De Morgan to William Rowan Hamilton, 1 May 1854, in Graves, *William Rowan Hamilton* 3:478.

132. Augustus De Morgan, Introductory lecture in mathematics, 1828, pp. 3–4, Ms Add.3, De Morgan Papers, Special Collections, University College Library, University College, London.

133. Ibid., 43.

134. Ibid., 37–40.

135. Augustus De Morgan to William Whewell, 30 April 1844, Add. Ms A.202100, Whewell Papers, Trinity College Library, Trinity College, Cambridge.

136. De Morgan, *Formal Logic* 171. Emphasis in the original.

137. Ibid.

138. Ibid., 180.

139. Ibid., 180–181.

140. Augustus De Morgan to William Whewell, 8 January 1851, Add. Ms A.202123, Whewell Papers, Trinity College Library, Trinity College, Cambridge.

141. Augustus De Morgan to William Rowan Hamilton, 28 April 1854, in Graves, *William Rowan Hamilton* 3:479.

142. De Morgan noted to William Rowan Hamilton that he had "especially looked at the psychology of inventive mathematicians." Augustus De Morgan to William Rowan Hamilton, 29 April 1855, in Graves, *William Rowan Hamilton* 3:495.

143. Augustus De Morgan to William Whewell, 1 April 1863, Add. Ms A.202147, Whewell Papers, Trinity College Library, Trinity College, Cambridge. Emphasis in the original.

144. Augustus De Morgan to William Whewell, 11 April 1863, Add. Ms A.202150, Whewell Papers, Trinity College Library, Trinity College, Cambridge.

145. Augustus De Morgan to William Rowan Hamilton, 6 December 1857, in Graves, *William Rowan Hamilton* 3:533.

146. Augustus De Morgan, Notes for an "Introductory Lecture," Ms Add. 3, De Morgan Papers, Special Collections, University College Library, University College, London.

147. Ibid., 6.

148. Ibid., 5.

149. Ibid., 11–12.

CHAPTER FIVE Earthly Calculations

1. See Gray, "The Nineteenth-Century Revolution in Mathematical Ontology," 226–248.

2. The classic treatment is Klein, *Vorträge Über Ausgewählte Fragen Der Elementargeometrie;* the revised and updated English translation of Klein's work, by Wooster Woodruff Beman, David Eugene Smith, and Raymond Clare Archibald, *Famous*

Problems of Elementary Geometry contains additional notes on the problem and its history.

3. See Lambert, *Beyträge zum Gebrauch der Mathematik und deren Anwendung*.

4. Lindemann, "Über die Zahl π," 213–225.

5. Rutherford, "Computation of the Ratio of the Diameter of a Circle to its Circumference to 208 Places of Figures," 281–283 (only the first 152 places were correct); Rutherford, "On the Extension of the Value of the Ratio of the Circumference of a Circle to its Diameter," 273–275.

6. Shanks, *Rectification of the Circle*. For a review of the π milestones, see Glaisher, "Remarks on the Calculation of π," 119–128, esp. table on 122. Shanks's *Rectification of the Circle* was purchased by William Whewell, Augustus De Morgan, and George Biddell Airy, among others (Shanks, *Rectification of the Circle* ix.).

7. Parker, *Quadrature of the Circle* (originally published in 1851).

8. Ibid., 95; see also 14, 21.

9. Ibid., 96, 100. Emphasis in the original.

10. Ibid., 39, 130.

11. See Schepler, "The Chronology of Pi," 216–228, 279–283. The phenomenon was not restricted to the English-speaking world, although there seem to have been many more attempts in Britain and America than on the Continent.

12. Augustus De Morgan, "Quadrature of the Circle," *English Cyclopædia* London: Bradbury & Evans, 1856–1852, 5:874.

13. Townsend, *Men of the Time* 741–742.

14. Smith, *The Problem of Squaring the Circle Solved* 13, 16–17.

15. De Morgan, Review of *The Problem of Squaring the Circle Solved*, by James Smith, 319.

16. Smith, *The Quadrature of the Circle*.

17. De Morgan, Review of *The Quadrature of the Circle*, 627.

18. Ibid.

19. *The Athenaeum* 8 June 1861, 764.

20. Smith, *A Nut to Crack for the Readers of Prof. De Morgan's 'Budget of Paradoxes'*.

21. See, e.g., Smith, *The Quadrature of the Circle*, and Smith, *The Problem of Squaring the Circle Solved* iii, 5, 10–13.

22. George Biddell Airy, "Address," in *Report of the Thirty-First Meeting of the British Association for the Advancement of Science*. "Notices and Abstracts" appendix, 1–2.

23. This was almost an exact replay of earlier exclusions of "marginal sciences" from the British Association for the Advancement of Science. For instance, at the 1834 BAAS meeting at Edinburgh the prominent phrenologist George Combe asked to present his method. He was rejected first by the local secretary and then by Adam Sedgwick. Robert Graham, the chair of the natural history section, verbally abused Combe and phrenology in general, vowing never to let them into his section. Humiliated, the phrenologists had to set up their own yearly convention. The BAAS

elite took issue with phrenology for essentially the same reasons as mathematicians objected to circle squaring in 1861—they appeared to be disruptive and divisive, a threat to the stability and civility of the organization. See Morrell and Thackray, *Gentlemen of Science* 276–280.

24. Van Der Weyde, *The Quadrature of the Circle* 2, 14.

25. Ibid., 10. Note that Van Der Weyde followed the lead of Augustus De Morgan in tone and content to such a great extent that he ended up plagiarizing long sections of his British counterpart's attack on circle squarers. See, e.g., *The Quadrature of the Circle* 40, apparently copied from De Morgan's article "Quadrature of the Circle," *English Cyclopædia* London: Bradbury & Evans, 1856–1852, 5:874.

26. Parker, *Quadrature of the Circle* 65. Emphasis in the original.

27. Ibid., 28.

28. Ibid., 69.

29. Ibid., 5–6, 85.

30. Ibid., 85–86. Emphasis in the original.

31. De Morgan, *A Budget of Paradoxes* 472. A reprint of a letter dated 31 December 1867.

32. Ibid., 317–318, 332.

33. Carrick was the family name of the descendants of Robert Bruce, and there is no record of an Alick from the nineteenth century. In a fantastic and somewhat overwrought story in the preface, Playfair recounts his meeting with the mysterious Carrick, who entrusted his solution to the ancient geometrical problem to the young physician just before his death. Carrick, *The Secret of the Circle* 5–15.

34. Moore, *Geometrical Science*. For this lower-caliber type of circle-squaring effort, see also Fuller, *The Square of the Circle;* Willmon, *The Secret of the Circle and the Square;* Morrell, *The Squaration of the Circle;* Heisel, *Behold! The Grand Problem No Longer Unsolved*.

35. London Mathematical Society, *List of Members, 13th November, 1884*.

36. *Proceedings of the Edinburgh Mathematical Society* 1 (1883): 1–2.

37. New York Mathematical Society, *New York Mathematical Society List of Members*.

38. *Bulletin of the American Mathematical Society* 1 (1894–1895): 1.

39. Ball, *A History of Mathematics at Cambridge* 200–203; quoted in Howson, *A History of Mathematics Education in England* 212–214.

40. *The Gentleman's Diary* 1 (1741–1750): title page.

41. *The Lady's Diary* 113 (1816): title page.

42. See Parshall and Rowe, *The Emergence of the American Mathematical Research Community* 44, 51; Cajori, *The Teaching and History of Mathematics in the United States* 94–97. *The Mathematical Diary* was published from 1825 to 1832, *The Mathematical Companion* from 1828 to 1831.

43. See Richards, *Mathematical Visions* 40–41, and Becher, "William Whewell and Cambridge Mathematics," 1–48.

44. Until the end of the century other British universities had negligible numbers of honors graduates in mathematics. See the chart between pages 68 and 69 in Chapman, "University Training of Mathematicians," 61–70.

45. *Quarterly Journal of Pure and Applied Mathematics* 1 (1857): iii.

46. Ibid., iv.

47. Ibid.

48. *The Oxford, Cambridge and Dublin Messenger of Mathematics* 1 (1861–2): 1.

49. Ibid., 4.

50. [J. D. Runkle], introduction, *The Mathematical Monthly* 1 (1859): ii.

51. *American Journal of Mathematics* 1 (1878): iv. Emphasis in the original.

52. *The American Mathematical Monthly* 1 (1894): 1.

53. Ibid., 2.

54. Ibid., 1.

55. Biblical literalists never objected to the advanced mathematical conclusions about π. Although the Bible says far less about π than about other truths that modern science has contested, it does appear to claim that its value is 3, or even slightly less than 3. The critical passage is I Kings 7:23: "Then he made the molten sea; it was round, ten cubits from brim to brim, and five cubits high. A line of thirty cubits would encircle it completely" (repeated in II Chronicles 4:2). In other words, ten cubits was the diameter of the round bath, and thirty (or less) was supposedly the circumference; thus $\pi = 30/10 = 3$. It is difficult to argue that this was an approximation. The passage is part of a fairly specific description of holy items in the temple; other measurements do not seem to be rounded so broadly.

56. From the Rhind Papyrus, British Museum, the oldest known reference to π.

57. *The Problem of Squaring the Circle Solved* was even translated into French: *Quadrature du Cercle*, trans. Armand Granges (Bordeaux: Lafargue, 1863).

58. Smith, *The Problem of Squaring the Circle Solved* 32.

59. Ibid., 33–36.

60. Ibid., 32.

61. Parker, *Quadrature of the Circle* 119.

62. Ibid., 64. Emphasis in the original.

63. Ibid., 83. Emphasis in the original.

64. Myers, *The Quadrature of the Circle, the Square Root of Two, and the Right-Angled Triangle* vii.

65. Ibid.

66. Ibid., iv, vii–viii. Emphasis in the original.

67. Fuller, *The Square of the Circle* 31.

68. Babbage, *The Works of Charles Babbage* 9:40–44, 73–80. The *Ninth Bridgewater Treatise* was not actually an official volume in the Bridgewater series, which was established to counter the scientific undermining of theology.

69. Ibid., 73.

70. Ibid., 5–8.

71. Gregory, *Letters on the Evidences, Doctrines, and Duties, of the Christian Religion* 46.

72. Ibid., 56–59, quotation 59–60. Emphasis in the original.

73. Ibid., 60.

74. Byrne, *The Creed of Saint Athanasius Proved by Mathematical Parallel* vii.

75. Ibid., 3.

76. Ibid., 1.

77. De Morgan, *A Budget of Paradoxes* 296.

78. Quoted in De Morgan, *Budget of Paradoxes* 294. Many of these combinations of mathematics and theology, including *The Two Estates,* obviously echoed the religious use of infinity Blaise Pascal pioneered in his famous "wager" on God (*Pensées* 1670). See Rescher, *Pascal's Wager.*

79. Ware, *The Works of Henry Ware Jr., D.D.* 114–115.

80. Gregg, *Novum Organum Moralium.*

81. Ibid., 5–8, 17–19.

82. Ibid., 16.

83. Ibid., xxiii.

84. Ibid., 32.

85. Webb, "The Limits of Religious Liberty," 120–149.

86. Grote was consistently adamant on this point: five years after the Martineau affair, his last official act at UCL was to endow the Chair of Philosophy of Mind and Logic with the explicit condition that if "any such Minister should at any time or times be appointed by the Council to the Professorship . . . I direct that no payment should be made to him out of the present endowment." Harte and North, *The World of UCL* 73.

87. For a more detailed account of the Martineau affair, including the text of De Morgan's letter of resignation, see S. De Morgan, *Memoir* 337–361, 369, 373–376, and Carpenter, *James Martineau* 432–433.

88. Augustus De Morgan to George Biddell Airy, 6 November 1866, Ms RGO 6/ 381, f. 183, Royal Greenwich Observatory Papers, Special Collections, Cambridge University Library, Cambridge.

89. Augustus De Morgan to William Heald, 20 August 1867, in S. De Morgan, *Memoir* 73.

90. Augustus De Morgan, Open letter to his students, 9 May 1867, Ms 913, Box 2, Packet 11, De Morgan Papers, Special Collections, University of London Library, London. Emphasis in the original.

91. S. De Morgan, *Memoir* 336, 393.

92. Augustus De Morgan to James Martineau, 19 December 1869, in S. De Morgan, *Memoir* 398.

93. S. De Morgan, *Memoir* 365.

94. Ibid., 385.

95. On other attempts by academics to establish unity in a divided age, see Harvie, *The Lights of Liberalism.*

96. Venn, *Logic of Chance* 434.

97. Ibid., 424.

98. Venn, *Symbolic Logic* xviii–xxvii.

99. De Morgan, *A Budget of Paradoxes* 199, 295, 298–300, 394–395, 436–437.

100. Ibid., 296.

101. Ibid., 199.

102. Ibid., 251.

103. Dirk J. Struik, "A Story Concerning Euler and Diderot," *Isis* 31 (1940): 431–432.

104. Ibid.

105. De Morgan, *A Budget of Paradoxes* 473–474.

106. Jevons, *The Principles of Science* 765.

107. Ibid., 43.

108. Ibid., 151.

109. Ibid., 766–768.

110. Kline, *Mathematics: The Loss of Certainty* 172–179.

111. Cantor, *Mathematische Annalen* 563–564; quoted in Kline, *Mathematical Thought from Ancient to Modern Times* 1031.

112. See Richards, *Mathematical Visions* esp. 13–59; Kline, *Mathematical Thought from Ancient to Modern Times* 1031–1038; and Torretti, *Philosophy of Geometry from Riemann to Poincaré*.

113. See Frederick Pollock's biographical introduction to Clifford, *Lectures and Essays* 1:1–55, esp. 3, 31–52.; Stephen, "William Kingdon Clifford," 4:538–541.

114. See Clifford's letter to Frederick Pollock in Clifford, *Lectures and Essays* 1:45–46.

115. For an outline of this proposed book, see ibid., 1:71–72.

116. See Clifford, "On the Aims and Instruments of Scientific Thought," 499–512; Richards, *Mathematical Visions* 109–114.

117. [Stewart and Tait], *The Unseen Universe; or Physical Speculations on a Future State*.

118. The authors had revealed their intentions in a simple cryptogram of their names in *Nature* on 15 October 1874. See Knott, *Life and Scientific Work of Peter Guthrie Tait* 236–242. The first edition to indicate their true identities was the fourth.

119. Clifford, Review of *The Unseen Universe*, 785.

120. Ibid., 787.

121. Ibid., 787–788.

122. Ibid., 791. For more on the reaction to *The Unseen Universe*, see Heimann, "*The Unseen Universe*: Physics and the Philosophy of Nature in Victorian Britain," 73–79, and contemporary reviews in *Nature* 12 (1875): 41–43; *The London Quarterly Review* 45 (1875–1876): 49–83; [William Mitchinson Hicks], *The British Quarterly Review* 64 (1876): 35–57; E[dward] C[aird], *Fraser's Magazine* 93 (1876): 60–68; *The Friend*, 2 October 1876, 252–253; *The Nation*, 27 May 1875, 366–367.

123. See, e.g., Blackwell, *The Physical Basis of Immortality* esp. 10–11, 74–75; for a later imitation, see Leighton, *Intimations of Eternal Life* esp. 44, 58.

124. Peter Guthrie Tait to Rev. Paton, 18 January 1883, Ms Gen.2169, f. 180, Special Collections, University of Edinburgh Library, University of Edinburgh, Edinburgh. Emphasis in the original.

125. See Peter Guthrie Tait, "Religion and Science," *The Scots Observer*, 8 December 1888, reprinted in Knott, *Life and Scientific Work of Peter Guthrie Tait* 293–295.

126. Mrs. Laurence Humphry, "Notes and Recollections," in Stokes, *Memoir and Scientific Correspondence of the Late Sir George Gabriel Stokes* 1–49; Lord Rayleigh, "Obituary Notice," in Stokes, *Mathematical and Physical Papers* ix–xxv.

127. Early on the Gifford lectureship conferred more wealth than honor. As Otto Pfleiderer, the German theologian who succeeded Stokes as the Gifford lecturer famously commented, "The honor is not great but the honorarium is colossal." Jaki, *Lord Gifford and his Lectures* 11.

128. George Gabriel Stokes to Peter Guthrie Tait, 15 September 1890, Ms Gen.2169, f. 175, Special Collections, University of Edinburgh Library, University of Edinburgh, Edinburgh.

129. Stokes, *Natural Theology*.

130. George Gabriel Stokes to Peter Guthrie Tait, 15 September 1890, Ms Gen.2169, f. 175, Special Collections, University of Edinburgh Library, University of Edinburgh, Edinburgh.

131. Sylvester, "Address on Commemoration Day at Johns Hopkins University, 22 February, 1877," in *The Collected Mathematical Papers of James Joseph Sylvester* 3:81.

132. Ibid., 80, 82.

133. Baker, "Biographical Notice," in *The Collected Mathematical Papers of James Joseph Sylvester* 4:xv–xxxvii; *Dictionary of National Biography* 19:258–260.

134. Sylvester, "A Probationary Lecture on Geometry," in *The Collected Mathematical Papers of James Joseph Sylvester* 2:7, 9.

135. Sylvester, "Presidential Address to Section 'A' of the British Association," in *The Collected Mathematical Papers of James Joseph Sylvester* 2:659.

136. Sylvester, "On the Theory of the Syzygetic Relations of Two Rational Integral Functions," 407.

137. Sylvester, "Lectures on the Theory of Reciprocants," in *The Collected Mathematical Papers of James Joseph Sylvester* 4:329.

138. Fraser, Review of *The Common Sense of the Exact Sciences*, by William Kingdon Clifford, 439.

139. Ibid.

140. Late-twentieth-century biographies of Dodgson clearly overemphasize (possibly anachronistically) his psychological quirks and understate his involvement with mathematics. See, for example, Thomas, *Lewis Carroll* and Bloom, *Lewis Carroll: Modern Critical Views,* both of which barely mention his education and employment in mathematics. (Significantly, most recent biographies call him Lewis Carroll, not Charles Dodgson, the name he generally used for his mathematical works.) In perhaps the most tenous interpretation, one biographer concludes before writing off

Dodgson's mathematical corpus that "it is highly likely that the motivation behind his publishing of works in mathematics was to allay possible criticism that he was using the position of don inappropriately." Wallace, *The Agony of Lewis Carroll* 180.

141. Warren Weaver, "Lewis Carroll: Mathematician," *Scientific American* 194 (April 1956): 116–128.

142. Dodgson, *A Syllabus of Plane Algebraical Geometry*. He also peppered *Alice's Adventures in Wonderland* (1865) and *Through a Looking Glass* (1871) with several mathematical problems; see Willerding, "Mathematics Through a Looking Glass," 209–219.

143. Dodgson, *The Complete Works of Lewis Carroll* 807–808; quoted in Engel, *From Clergyman to Don* 150.

144. Dodgson, *Euclid and his Modern Rivals*, 2d ed. x. Quotation is from the preface to the first edition.

145. Helena Pycior, "At the Intersection of Mathematics and Humor," 162.

146. Dodgson, *Euclid and his Modern Rivals* x.

147. Dodgson, *The Mathematical Pamphlets of Charles Lutwidge Dodgson and Related Pieces* 118–119.

148. Dodgson, "Simple Facts about Circle-Squaring," in *The Mathematical Pamphlets* 144–147.

149. Quoted by Francine F. Abeles in her introduction to Dodgson, *The Mathematical Pamphlets* 1.

150. See Dodgson, "Response to 'Infinitesimal or Zero?'" and "Something or Nothing?" in *The Mathematical Pamphlets* 213–220.

151. Dodgson, *Curiosa Mathematica*. Part I: *A New Theory of Parallels*, xiv.

152. Dodgson, *Euclid and his Modern Rivals* xi.

153. Francine F. Abeles, introduction to Dodgson, *The Mathematical Pamphlets* 3–4.

154. London Mathematical Society, *List of Members of the London Mathematical Society, 13th November, 1884*. Of 181 members, 18 were clergymen.

155. Dodgson, *The Letters of Lewis Carroll* 1:602–603. Emphasis in the original.

156. Ibid., 565. Many years earlier, Dodgson had written that a Cambridge mathematical education "dried up" men, while an Oxford mathematical education left their "imagination" intact (Dodgson, *The Letters of Lewis Carroll* 1:85).

157. Francine F. Abeles, introduction to Dodgson, *The Mathematical Pamphlets* 2–3.

158. Dodgson, *An Elementary Treatise on Determinants*.

159. Carroll, *A Tangled Tale* preface (unnumbered page).

160. Carroll, *The Game of Logic* preface (unnumbered page).

161. Dodgson, *The Complete Works of Lewis Carroll* 1118. Emphasis in the original.

162. As A. J. Engel notes, Dodgson was not alone in this retreat. Many others who expressed resistance to professionalization and research found themselves alienated in late Victorian Oxford. See Engel, *From Clergyman to Don* esp. 257–285.

163. See Fiske, "Mathematical Progress in America," 238–246; Parshall and Rowe, *The Emergence of the American Mathematical Research Community* 189–223.

164. Klein famously said that only ten to fifteen percent of all mathematics students were worthy for advanced training (Parshall and Rowe, *The Emergence of the American Mathematical Research Community* 190, n. 2).

165. Lobachevsky, *Geometrical Researches on the Theory of Parallels;* Bolyai, *The Science Absolute of Space.* Also important was Halsted's "Bibliography of Hyper-Space and Non-Euclidean Geometry."

166. Whitehead, *Universal Algebra* 4; quoted in Kline, *Mathematical Thought from Ancient to Modern Times* 1031.

167. See Griffin, *Russell's Idealist Apprenticeship;* Hylton, *Russell, Idealism, and the Emergence of Analytic Philosophy;* Jager, *The Development of Bertrand Russell's Philosophy;* Andersson, *In Quest of Certainty;* Levy, *Moore: G. E. Moore and the Cambridge Apostles;* Eames, *Bertrand Russell's Dialogue with His Contemporaries.*

168. Russell, *The Autobiography of Bertrand Russell* 38. This may be hindsight wisdom.

169. See Russell's laudatory preface to Clifford, *The Common Sense of the Exact Sciences.*

170. See Mill, *Autobiography;* Russell, *Collected Papers* 1:56–57.

171. Russell, *Collected Papers* 1:52.

172. Bradley, *Essays on Truth and Reality* 230–231.

173. Russell, *Collected Papers* 1:196.

174. Quoted in Dickinson, *J. McT. E. McTaggart* 38.

175. See Russell's description of this conclusion in his essay "Mysticism and Logic" (1914), in *Mysticism and Logic* 1–32.

176. Bertrand Russell, *Essay on the Foundations of Geometry* 1.

177. Ibid., 97. See Andersson, *In Quest of Certainty* 118.

178. Bertrand Russell, "Recent Work on the Principles of Mathematics," *International Monthly* 4 (1901): 83.

179. Ibid., 89.

180. Ibid., 83–4.

181. Ibid., 86.

182. Ibid., 92–5.

183. Ibid., 98.

184. Ibid., 99.

185. Ibid., 100. Russell called Euclid's *Elements* "terribly long-winded" and thought it "nothing less than a scandal" that it was still taught in schools (ibid.).

186. Ibid., 93.

187. Ibid., 101.

188. Ibid., 84.

189. Ibid.

Bibliography

Unpublished Sources

Boole, George. Papers. Library of the Royal Irish Academy, Dublin.

Boole, George. Papers. Library of the Royal Society of London, London.

Boole, George. Papers. Special Collections, University College Library, University College, Cork.

Brougham, Lord [Henry]. Papers. Special Collections, University College Library, University College, London.

Clifford, William Kingdon. Papers. Special Collections, University College Library, University College, London.

College Correspondence. Special Collections, University College Library, University College, London.

De Morgan, Augustus. Additional Manuscripts. Special Collections, British Library, London.

De Morgan, Augustus. Additional Manuscripts. Special Collections, National Library of Scotland, Edinburgh.

De Morgan, Augustus. Papers. Special Collections, University College Library, University College, London.

De Morgan, Augustus. Papers. Special Collections, University of London Library, University of London, London.

Hamilton, William Rowan. Papers. Special Collections, Trinity College Library, Trinity College, Dublin.

Herschel, John Frederick William. Papers. Library of the Royal Society of London, London.

Local History Collection. Lincolnshire Central Library, Lincoln.

Peirce, Benjamin. Papers. Houghton Library, Harvard University, Cambridge.

Peirce, Charles Sanders. Papers. Houghton Library, Harvard University, Cambridge.

Royal Greenwich Observatory. Papers. Special Collections, Cambridge University Library, Cambridge University, Cambridge.

Stokes, George Gabriel. Papers. Special Collections, University of Edinburgh Library, University of Edinburgh, Edinburgh.

Tait, Peter Guthrie. Papers. Special Collections, University of Edinburgh Library, University of Edinburgh, Edinburgh.

Whewell, William. Papers. Trinity College Library, Trinity College, Cambridge.

Printed Sources

Agassiz, Louis. *Contributions to the Natural History of the United States.* Boston: Little, Brown, 1857–1862.

Ahlstrom, Sydney. *A Religious History of the American People.* New Haven, CT: Yale University Press, 1972.

———. *Autobiography of Sir George Biddell Airy,* edited by Wilfred Airy. Cambridge: Cambridge University Press, 1896.

Akenside, Mark. *Poems.* London: W. Bowyer and J. Nichols, 1772.

Allen, Joseph Henry. *Sequel to 'Our Liberal Movement.'* Boston: Roberts Brothers, 1897.

Andersson, Stefan. *In Quest of Certainty: Bertrand Russell's Search for Certainty in Religion and Mathematics up to The Principles of Mathematics.* Stockholm: Almqvist & Wiksell, 1994.

Angherà, D. *Quadratura del Cerchio.* Malta, 1856.

Archibald, Raymond Clare, ed. *Benjamin Peirce: 1809–1880.* Oberlin, OH: Oberlin Press, 1925.

Aristotle. *Metaphysics.* Translated by Hugh Tredennick. London: William Heinemann, 1968.

Aspray, William and Philip Kitcher, eds. *History and Philosophy of Modern Mathematics.* Minneapolis: University of Minnesota Press, 1988.

Babbage, Charles. *The Ninth Bridgewater Treatise. A Fragment.* London: J. Murray, 1837.

———. *Passages from the Life of a Philosopher.* London: Longman, Green, Longman, Roberts & Green, 1864.

———. *The Works of Charles Babbage,* edited by Martin Campbell-Kelly. London: William Pickering, 1989.

Ball, Walter William Rouse. *A History of Mathematics at Cambridge.* Cambridge: Cambridge University Press, 1889.

Barrow, Logie. *Independent Spirits: Spiritualism and English Plebeians, 1850–1910.* London: Routledge & Kegan Paul, 1986.

Bartol, Cyrus Augustus. "The New Planet." *The Monthly Religious Magazine* 4 (1847): 77.

Becher, Harvey W. "William Whewell and Cambridge Mathematics." *Historical Studies in the Physical Sciences* 11 (1980): 1–48.

———. "Woodhouse, Babbage, Peacock and Modern Algebra." *Historia Mathematica* 7 (1980): 389–400.

Bell, Eric Temple. *The Development of Mathematics,* 2nd ed. New York: McGraw-Hill, 1945.

Bellot, Hugh Hale. *University College London, 1826–1926*. London: University of London Press, 1929.

Berman, Morris. "'Hegemony' and the Amateur Tradition in British Science." *Journal of Social History* 8 (1974/1975): 30–50.

———. *Social Change and Scientific Organization: The Royal Institution, 1799–1844*. Ithaca, NY: Cornell University Press, 1978.

Blackwell, Antoinette Brown. *The Physical Basis of Immortality*. New York: G. P. Putnam's Sons, 1876.

Bloom, Harold, ed. *Lewis Carroll: Modern Critical Views*. New York: Chelsea House, 1987.

Bochenski, I. M. *A History of Formal Logic*. Translated by Ivo Thomas. Notre Dame, IN: University of Notre Dame Press, 1961.

Bode, Johann Elert. *Anleitung zur Kenntniss des gestirnen Himmels*. Hamburg, 1772.

Bolyai, János. *The Science Absolute of Space*. Translated by George Bruce Halsted. Austin: The University of Texas Press, 1896.

Boole, George. *An Address on the Genius and Discovery of Isaac Newton*. Lincoln: Printed at the Gazette Office, 1835.

———. "Analytical Geometry." *Cambridge Mathematical Journal* 2 (1840–1841): 179–188.

———. "The Calculus of Logic." *Cambridge and Dublin Mathematical Journal* 3 (1848): 183–198.

———. *The Claims of Science, Especially as Founded in its Relations to Human Nature*. London, 1851.

———. "Exposition of a General Theory of Linear Transformations." *Cambridge Mathematical Journal* 3 (1841–1842): 1–20, 106–119.

———. *An Investigation of the Laws of Thought on Which are Founded the Mathematical Theories of Logic and Probabilities*. London: Walton and Maberly, 1854.

———. *The Mathematical Analysis of Logic, being an Essay towards a Calculus of Deductive Reasoning*. Cambridge: Macmillan, Barclay, & Macmillan, 1847.

———. "On a Certain Symbolical Equation." *Cambridge and Dublin Mathematical Journal* 2 (1847): 7–12.

———. "Notes on Quaternions." *Philosophical Magazine* 33 (1848): 278–280.

———. *The Right Use of Leisure*. London: J. Nisbet & Co., 1847.

———. *Selected Manuscripts on Logic and Its Philosophy*, edited by Ivor Grattan-Guinness and Gerard Bornet. Boston: Birkhauser, 1997.

———. *The Social Aspects of Intellectual Culture*. Cork: The Cuvierian Society, 1855.

———. *Studies in Logic and Probability by George Boole*, edited by Rush Rhees. London: Watts and Co., 1952.

———. *A Treatise on Differential Equations*. Cambridge: Macmillan, 1859.

———. *A Treatise on the Calculus of Finite Differences*. Cambridge: Macmillan, 1860.

Boole, Mary Everest. *Collected Works*. London: The C. W. Daniel Co., 1931.

Boyer, Carl B. *A History of Mathematics*. New York: John Wiley and Sons, 1968.

———. *The History of the Calculus and its Conceptual Development*. New York: Dover, 1949.

Bradley, F. H. *Essays on Truth and Reality*. Oxford: Oxford University Press, 1914.

———. *The Principles of Logic*. London: Oxford University Press, 1922.

Brewster, David. "Researches Respecting the New Planet Neptune." *North British Review* 7 (May 1847): 111.

British Association for the Advancement of Science. *Report of the Seventeenth Meeting of the British Association for the Advancement of Science*. London: John Murray, 1848.

British Association for the Advancement of Science. *Report of the Thirty-First Meeting of the British Association for the Advancement of Science*. London: John Murray, 1862.

Brock, W. H. "Geometry and the Universities: Euclid and his Modern Rivals." *History of Education* 4 (1975): 21–35.

Brower, William. *The Quadrature of the Circle*. Philadelphia: Sower, Potts & Co., 1874.

Bruce, Robert V. *The Launching of Modern American Science: 1846–1876*. New York: Alfred A. Knopf, 1987.

Bushnell, Horace. *An Oration Delivered Before the Society of Phi Beta Kappa, at Cambridge*. Cambridge: G. Nichols, 1848.

Butler, Jon. *Awash in a Sea of Faith: Christianizing the American People*. Cambridge, MA: Harvard University Press, 1990.

Byrne, Oliver. *The Creed of Saint Athanasius Proved by Mathematical Parallel*. London: William Day, 1839.

Caird, Edward. Review of *The Unseen Universe*, by Balfour Stewart and Peter Guthrie Tait. *Fraser's Magazine* 93 (1876): 60–68.

Cajori, Florian. *A History of Mathematics*. New York: Macmillan, 1922.

———. *Mathematics in Liberal Education: A Critical Examination of the Judgments of Prominent Men of the Ages*. Boston: Christopher Publishing House, 1928.

———. *The Teaching and History of Mathematics in the United States*. Washington, DC: Government Printing Office, 1890.

Cantor, Geoffrey. *Michael Faraday: Sandemanian and Scientist*. London: Macmillan, 1991.

Carpenter, J. Estlin. *James Martineau*. London: Philip Green, 1905.

Carrick, Alick. *The Secret of the Circle*. London: Henry Sotheran and Co., 1876.

Carroll, Lewis. [Charles Lutwidge Dodgson]. *The Game of Logic*. London: Macmillan, 1887.

———. *A Tangled Tale*. London: Macmillan, 1885.

Cashdollar, Charles D. *The Transformation of Theology, 1830–1890: Positivism and Protestant Thought in Britain and America*. Princeton, NJ: Princeton University Press, 1989.

Cayley, Arthur. *The Collected Mathematical Papers of Arthur Cayley*. Cambridge: Cambridge University Press, 1889–1897.

Chadwick, Owen. *The Victorian Church*. London: A. & C. Black, 1966–1970.

Chapman, S. "University Training of Mathematicians." *The Mathematical Gazette* 30 (1946): 61–70.

Christmas, Henry. *Echoes of the Universe: From the World of Matter and the World of Spirit*. Philadelphia: A. Hart, 1850.

Clarke, James Freeman. *Autobiography, Diary, and Correspondence*, edited by Edward Everett Hale. Boston: Houghton Mifflin, 1891.

Clifford, William Kingdon. *The Common Sense of the Exact Sciences*, edited by Karl Pearson. Preface by Bertrand Russell. New York: A. A. Knopf, 1946.

———. *Lectures and Essays*, edited by Leslie Stephen and Frederick Pollock. London: Macmillan, 1879.

———. *Mathematical Papers*, edited by Robert Tucker. London: Macmillan, 1882.

———. "On the Aims and Instruments of Scientific Thought." *Macmillan's Magazine* 26 (1872): 499–512.

———. "On the Space-Theory of Matter." *Proceedings of the Cambridge Philosophical Society* 2 (1864–1876): 157–158.

———. Review of *The Unseen Universe*, by Balfour Stewart and Peter Guthrie Tait. *The Fortnightly Review* 23 (1875): 776–793.

Comte, Auguste. *The Philosophy of Mathematics*. Translated from *Cours de Philosophie Positive* by W. M. Gillespie. New York: Harper & Brothers, 1851.

Corcoran, John, ed. *Ancient Logic and its Modern Interpretations*. Dordrecht, The Netherlands: D. Reidel Publishing Co., 1974.

Corsi, Pietro. *Science and Religion: Baden Powell and the Anglican Debate, 1800–1860*. Cambridge: Cambridge University Press, 1987.

Coleridge, Samuel Taylor. *Aids to Reflection*, 2nd ed. London: Hurst, Chance, and Co., 1831.

———. *Biographia Literaria*. Princeton, NJ: Princeton University Press, 1983.

———. *Collected Letters*, edited by E. L. Griggs. Oxford: Oxford University Press, 1956–1971.

———. *The Collected Works of Samuel Taylor Coleridge*, edited by Kathleen Coburn. London: Routledge & Kegan Paul, 1969–1992.

———. *Confessions of an Inquiring Spirit*, edited by Henry Nelson Coleridge. London: William Pickering, 1840.

———. *Logic*. Princeton, NJ: Princeton University Press, 1981.

———. *The Notebooks of Samuel Taylor Coleridge*. New York: Pantheon Books, 1957–1990.

Coolidge, Julian L. "The Story of Mathematics at Harvard." *Harvard Alumni Bulletin* 26 (January 1924): 372–378.

Crowther, J. G. *Scientific Types*. London: Barrie & Rockliff/The Cresset Press, 1969.

Daniels, George H. "The Process of Professionalization in American Science: the Emergent Period, 1820–1860." *Isis* 58 (1967): 151–166.

Dauben, Joseph W. *The History of Mathematics from Antiquity to the Present: A Selective Bibliography*. New York: Garland Publishing, 1985.

Davie, George Elder. *The Democratic Intellect: Scotland and her Universities in the Nineteenth Century*. Edinburgh: University of Edinburgh Press, 1961.

De Morgan, Augustus. *Address to the London Mathematical Society*. London: William Clowes and Sons, 1865.

———. *A Budget of Paradoxes*. London: Longmans, Green, 1872.

———. *Formal Logic*. London: Taylor and Walton, 1847.

———. "On Divergent Series, and Various Points of Analysis Connected with Them." *Transactions of the Cambridge Philosophical Society* 8 (1849): 182–203.

———. "On Infinity; and on the Sign of Equality." *Transactions of the Cambridge Philosophical Society* 11 (1866): 145–189.

———. "On the Foundation of Algebra." *Transactions of the Cambridge Philosophical Society* 7 (1842): 173–187.

———. "On the Foundation of Algebra, No. II." *Transactions of the Cambridge Philosophical Society* 7 (1842): 287–300.

———. "On the Foundation of Algebra, No. III." *Transactions of the Cambridge Philosophical Society* 8 (1849): 141–142.

———. "On the Foundation of Algebra, No. IV." *Transactions of the Cambridge Philosophical Society* 8 (1849): 241–254.

———. "On the Signs + and − in Geometry (continued), and On the Interpretation of the Equation of a Curve." *The Cambridge and Dublin Mathematical Journal* 7 (1852): 242–249.

———. *On the Study and Difficulty of Mathematics*, 3rd ed. Chicago: Open Court Publishing, 1910.

———. *On the Syllogism and Other Logical Writings*, edited by Peter Heath. New Haven, CT: Yale University Press, 1966.

———. Review of *The Problem of Squaring the Circle Solved*, by James Smith. *The Athenaeum*, 5 March 1859, 319.

———. Review of *The Quadrature of the Circle: Correspondence between an Eminent Mathematician and James Smith, Esq.*, by James Smith. *The Athenaeum*, 11 May 1861, 627.

———. *Trigonometry and Double Algebra*. London: Taylor, Walton, and Maberly, 1849.

De Morgan, Sophia Elizabeth. *Memoir of Augustus De Morgan*. London: Longmans, Green, 1882.

———. *From Matter to Spirit*. Preface by Augustus De Morgan. London: Longmans, Green, 1863.

Desmond, Adrian. *The Politics of Evolution: Morphology, Medicine, and Reform in Radical London*. Chicago: University of Chicago Press, 1989.

Dickinson, G. Lowes. *J. McT. E. McTaggart*. Cambridge: Cambridge University Press, 1931.

The Dictionary of National Biography Oxford: Oxford University Press, 1917.

Dirks, John E. *The Critical Theology of Theodore Parker*. New York: Columbia University Press, 1948.

Dodgson, Charles Lutwidge. *The Complete Works of Lewis Carroll*. London: The None-such Press, 1973.

———. *Curiosa Mathematica*. London: Macmillan, 1888.

———. *An Elementary Treatise on Determinants with Their Application to Simultaneous Linear Equations and Algebraical Geometry*. London: Macmillan, 1867.

———. *Euclid and his Modern Rivals*, 2nd ed. London: Macmillan, 1885.

———. *The Letters of Lewis Carroll*, edited by Morton N. Cohen. New York: Oxford University Press, 1979.

———. *The Mathematical Pamphlets of Charles Lutwidge Dodgson and Related Pieces*, edited by Francine F. Abeles. New York: The Lewis Carroll Society of North America, 1994.

———. *A Syllabus of Plane Algebraical Geometry, Systematically Arranged with Formal Definitions, Postulates, and Axioms*. Oxford: James Wright, 1860.

Doyle, Arthur Conan. *The History of Spiritualism*. New York: George H. Doran, 1926.

Dubbey, J. M. "Babbage, Peacock and Modern Algebra." *Historia Mathematica* 4 (1977): 295–302.

Duren, Peter, ed. *A Century of Mathematics in America*. Providence, RI: American Mathematical Society, 1988–1989.

Eames, Elizabeth Ramsden. *Bertrand Russell's Dialogue with His Contemporaries*. Carbondale, IL.: Southern Illinois University Press, 1989.

Elliot, Clark A., and Margaret W. Rossiter, eds. *Science at Harvard University*. Bethlehem, PA: Lehigh University Press, 1992.

Emerson, Edward Waldo. *The Early Years of the Saturday Club*. Boston: Houghton Mifflin, 1918.

Emerson, Ralph Waldo. *The Collected Works of Ralph Waldo Emerson*. Cambridge, MA: Harvard University Press, 1971.

———. *The Complete Essays and Other Writings of Ralph Waldo Emerson*. New York: Random House, 1940.

———. *Emerson in His Journals*, edited by Joel Porte. Cambridge, MA: Harvard University Press, 1982.

Engel, A. J. *From Clergyman to Don*. Oxford: Clarendon Press, 1983.

Enros, Philip. "The Analytical Society (1812–1813): Precursor of the Renewal of Cambridge Mathematics." *Historia Mathematica* 10 (1983): 24–47.

Euclid. *Elements*. Translated by H. Billingsley. London: John Daye, 1570.

Everett, Edward. *Orations and Speeches on Various Occasions*. Boston: Little, Brown, and Co., 1859.

Faraday, Michael. *Experimental Researches in Chemistry and Physics*. London: R. Taylor and W. Francis, 1859.

———. *The Selected Correspondence of Michael Faraday*, edited by L. Pearce Williams. Cambridge: Cambridge University Press, 1971.

Farrar, Adam S. *Sermons Preached in St. Mary's, Oxford, Before the University*. Philadelphia: Smith, English, & Co., 1860.

Ficino, Marsilio. *Three Books on Life,* edited by Carol V. Kaske and John R. Clark. Binghamton, NY: The Renaissance Society of America, 1989.

Fiske, Thomas. "Mathematical Progress in America." *Bulletin of the American Mathematical Society* 11 (1905): 238–246.

Fraser, A. Y. Review of *The Common Sense of the Exact Sciences,* by William Kingdon Clifford. *The Academy,* 20 June 1885, 439–440.

French, John C. *History of the University Founded by Johns Hopkins.* Baltimore: Johns Hopkins University Press, 1946.

French, Peter J. *John Dee: The World of an Elizabethan Magus.* London: Routledge & Kegan Paul, 1972.

Friedman, Michael. "Kant's Theory of Geometry." *Philosophical Review* 94 (1985): 455–506.

Foote, George A. "The Place of Science in the British Reform Movement, 1830–1850." *Isis* 42 (1951): 192–208.

Forsyth, A. R. "Old Tripos Days at Cambridge." *The Mathematical Gazette* 19 (1935): 162–179.

Fuller, Rufus. *The Square of the Circle.* Cambridge: The Riverside Press, 1893.

Galilei, Galileo. *Discoveries and Opinions of Galileo.* Translated by Stillman Drake. Garden City, NY: Doubleday, 1957.

Garland, Martha McMackin. "Newman in His Own Day." In *John Henry Newman, The Idea of a University,* edited by Frank M. Turner, 265–281. New Haven, CT: Yale University Press, 1996.

Gauld, Alan. *The Founders of Psychical Research.* New York: Schocken Books, 1968.

Gillies, Donald, ed. *Revolutions in Mathematics.* Oxford: Clarendon Press, 1992.

Glaisher, J. W. L. "Remarks on the Calculation of π." *The Messenger of Mathematics,* 2nd ser., 2 (1873): 119–128.

Grattan-Guiness, Ivar. *The Development of the Foundations of Mathematical Analysis from Euler to Riemann.* Cambridge, MA: MIT Press, 1970.

———. "Does History of Science Treat of the History of Science? The Case of Mathematics." *History of Science* 28 (1990): 149–173.

———. *The Search for Mathematical Roots, 1870–1940: Logics, Set Theories and the Foundations of Mathematics from Cantor through Russell to Gödel.* Princeton, NJ: Princeton University Press, 2000.

———. "University Mathematics at the Turn of the Century: Unpublished Recollections by W. H. Young." *Annals of Science* 28 (1973): 369–384.

Graves, Robert Perceval. *Life of Sir William Rowan Hamilton.* Dublin: Hodges, Figgis, & Co., 1885.

Gray, Jeremy. "The Nineteenth-Century Revolution in Mathematical Ontology." In *Revolutions in Mathematics,* edited by Donald Gillies, 226–248. Oxford: Clarendon Press, 1992.

Gregg, Tresham Dames. *Novum Organum Moralium.* London: H. Baillière, 1859.

Gregory, Olinthus. *Letters on the Evidences, Doctrines, and Duties, of the Christian Religion,* 9th ed. London: Henry G. Bohn, 1851.

Griffin, Nicholas. *Russell's Idealist Apprenticeship*. Oxford: The Clarendon Press, 1991.

Grobel, Monica C. "The Society for the Diffusion of Useful Knowledge 1826–1846 and its Relation to Adult Education in the First Half of the XIXth Century." Ph.D. dissertation, University of London, 1932.

Grosser, Morton. *The Discovery of Neptune*. Cambridge, MA: Harvard University Press, 1962.

Grosvenor, Cyrus Pitt. *The Circle Squared*. New York, 1868.

Guy, Jeff. *The Heretic: A Study of the Life of John William Colenso, 1814–1883*. Johannesburg: Raven Press, 1983.

Hailperin, Theodore. *Boole's Logic and Probability*, 2nd ed. Amsterdam: North-Holland, 1986.

Hales, E. E. Y. *The Catholic Church in the Modern World*. Garden City, NY: Hanover House, 1958.

Halsted, George Bruce. "Bibliography of Hyper-Space and Non-Euclidean Geometry." *American Journal of Mathematics* 1 (1878): 261–276, 385–395, and 2 (1879): 65–70.

———. "Light from Non-Euclidean Spaces on the Teaching of Elementary Geometry." *Scientiae Baccalaureus* 1 (1890): 255–260.

———. "Our Belief in Axioms, and the New Spaces." *Scientiae Baccalaureus* 1 (1890): 11–19.

Hamilton, William. *Discussions on Philosophy and Literature*, 2nd ed. New York: Harper & Brothers, 1858.

———. "On the Study of Mathematics." *Edinburgh Review* 126 (January 1836): 409–455.

———. *Philosophy of William Hamilton*. New York: D. Appleton, 1857.

Hankins, Thomas L. "Algebra as Pure Time: William Rowan Hamilton and the Foundations of Algebra." In *Motion and Time, Space and Matter: Interrelations in the History of Philosophy and Science*, edited by Peter K. Machamer and Robert G. Turnbull, 327–359. Columbus: Ohio State University Press, 1978.

———. *Sir William Rowan Hamilton*. Baltimore: The Johns Hopkins University Press, 1980.

———. "Triplets and Triads: Sir William Rowan Hamilton on the Metaphysics of Mathematics." *Isis* 68 (1977): 175–193.

Harley, Robert. *George Boole, F.R.S. An Essay, Biographical and Expository*. London: Benjamin Pardon, 1866.

Harte, Negley and John North. *The World of UCL, 1828–1990*. London: University College London, 1991.

Harvie, Christopher T. *The Lights of Liberalism: University Liberals and the Challenge of Democracy*. London: Allen Lane, 1976.

Hatch, Nathan O. *The Democratization of American Christianity*. New Haven, CT: Yale University Press, 1989.

Hawkins, Hugh. *Pioneer: A History of the Johns Hopkins University, 1874–1889*. Ithaca, NY: Cornell University Press, 1960.

Heimann, P. M. "The Unseen Universe: Physics and the Philosophy of Nature in Victorian Britain." *The British Journal for the History of Science* 6 (1972): 73–79.

Heisel, Carl Theodore. *Behold! The Grand Problem No Longer Unsolved*. Cleveland, OH: S. J. Monck, 1931.

Helmstadter, Richard, ed. *Freedom and Religion in the Nineteenth Century*. Stanford: Stanford University Press, 1997.

Herschel, John Frederick William. *Essays from the Edinburgh and Quarterly Reviews with Addresses and Other Pieces*. London: Longman, Brown, Green, Longmans, & Roberts, 1857.

———. *A Preliminary Discourse on the Study of Natural Philosophy*. London: Longmans, Rees, Orme, Brown, and Green, 1831.

Hesse, Mary B. "Boole's Philosophy of Logic." *Annals of Science* 8 (1952): 61–81.

Hicks, William Mitchinson. Review of *The Unseen Universe*, by Balfour Stewart and Peter Guthrie Tait. *The British Quarterly Review* 64 (1876): 35–57.

Hill, Thomas. *Annual Report of the President and Treasurer of Harvard College*. Cambridge, MA: Welch, Bigelow, and Co., 1864.

———. *Annual Report of the President and Treasurer of Harvard College*. Cambridge, MA: Welch, Bigelow, and Co., 1865.

———. "Benjamin Peirce." *The Harvard Register* 1 (1880): 91–92.

———. "Books That Have Helped Me." *Forum* 4 (1887–1888): 388–396.

———. *First Lessons in Geometry*, rev. ed. Boston: William Ware and Co., 1878.

———. *Geometry and Faith*. New York: C. S. Francis, 1849.

———. *Geometry and Faith*, 2nd ed. New York: G. P. Putnam's Sons, 1874.

———. *Geometry and Faith*, 3rd ed. Boston: Lee and Shepard, 1882.

———. *God Known by His Works*. Portland, ME: William M. Marks, 1887.

———. *Liberal Education, An Address Before the Phi Beta Kappa Society of Harvard College, July 22, 1858*. Cambridge, MA: John Bartlett, 1858.

———. *Philosophy Higher than Science*. Portland, ME: William M. Marks, 1876.

———. *Religion in Public Instruction: Baccalaureate Address Delivered Before the Graduating Class of Antioch College*. Boston: Little, Brown, and Co., 1860.

———. Review of *Logic*, by John Stuart Mill. *Christian Examiner* 40 (May 1846): 366–368.

———. *The Natural Sources of Theology*. Andover, MA: W. F. Draper, 1877.

———. *The True Order of Studies*, rev. ed. New York: G. P. Putnam's Sons, 1876.

Hoar, George F. "Harvard College Fifty-Eight Years Ago." *Scribner's Magazine* 28 (1900): 64.

Houghton, Walter E. *The Victorian Frame of Mind, 1830–1870*. New Haven, CT: Yale University Press, 1957.

Howson, Geoffrey. *A History of Mathematics Education in Modern England*. Cambridge: Cambridge University Press, 1982.

Hylton, Peter. *Russell, Idealism, and the Emergence of Analytic Philosophy*. Oxford: Clarendon Press, 1990.

Hyman, Anthony. *Charles Babbage: Pioneer of the Computer.* Princeton, NJ: Princeton University Press, 1982.

Inge, William Ralph. *The Platonic Tradition in English Religious Thought.* London: Longmans, Green, 1926.

Jager, Ronald. *The Development of Bertrand Russell's Philosophy.* London: Allen and Unwin, 1972.

Jaki, Stanley L. *Lord Gifford and his Lectures,* 2nd ed. Edinburgh: Scottish Academic Press, 1995.

James, Mary Ann. "Engineering an Environment for Change: Bigelow, Peirce, and Early Nineteenth-Century Practical Education at Harvard." In *Science at Harvard University,* edited by Clark A. Elliot and Margaret W. Rossiter, 55–75. Bethlehem, PA: Lehigh University Press, 1992.

Jevons, William Stanley. *Letters and Journal of W. Stanley Jevons.* London: Macmillan, 1886.

———. *The Principles of Science: A Treatise on Logic and Scientific Method.* London: Macmillan, 1874.

———. *Pure Logic and Other Minor Works,* edited by Robert Adamson and Harriet A. Jevons. London: Macmillan, 1890.

Johnston, Mark D. *The Spiritual Logic of Ramon Llull.* Oxford: Oxford University Press, 1987.

Jones, Henry Bence. *The Life and Letters of Faraday.* London: Longmans, Green, 1870.

Kant, Immanuel. *Critique of Pure Reason.* Translated by Norman Kemp Smith. London: Macmillan, 1933.

Kitcher, Philip. *The Nature of Mathematical Knowledge.* New York: Oxford University Press, 1983.

Klein, Felix. *Famous Problems of Elementary Geometry.* Translated and edited by Wooster Woodruff Beman, David Eugene Smith, and Raymond Clare Archibald. New York: G. E. Stechert & Co., 1930.

———. *Vorträge Über Ausgewählte Fragen Der Elementargeometrie.* Leipzig: B. G. Teubner, 1895.

Klein, Jacob. *Greek Mathematical Thought and the Origin of Algebra.* Cambridge, MA: MIT Press, 1968.

Kline, Morris. *Mathematical Thought from Ancient to Modern Times.* New York: Oxford University Press, 1972.

———. *Mathematics: The Loss of Certainty.* New York: Oxford University Press, 1980.

Kneale, William. "Boole and the Revival of Logic." *Mind* 57 (1948): 149–175.

Knight, Frida. *University Rebel: The Life of William Frend (1757–1841).* London: Victor Gollancz, 1971.

Knott, Cargill Gilston. *Life and Scientific Work of Peter Guthrie Tait.* Cambridge: Cambridge University Press, 1911.

Koppelman, Elaine. "The Calculus of Operations and the Rise of Abstract Algebra." *Archive for History of Exact Sciences* 8 (1971): 155–242.

Ladu, Arthur I. "Channing and Transcendentalism." *American Literature* 11 (1939): 129–137.

Laita, Luis M. "Boolean Algebra and its Extra-logical Sources: The Testimony of Mary Everest Boole." *History and Philosophy of Logic* 1 (1980): 37–60.

Lambert, Johann Heinrich. *Beyträge zum Gebrauch der Mathematik und deren Anwendung*. Berlin, 1765–1772.

Land, William G. *Thomas Hill*. Cambridge, MA: Harvard University Press, 1933.

Leighton, Caroline C. *Intimations of Eternal Life*. Boston: Lee and Shepard, 1891.

Levere, Trevor H. "Coleridge, Chemistry, and the Philosophy of Nature." *Studies in Romanticism* 16 (1977): 349–379.

———. "S. T. Coleridge: A Poet's View of Science." *Annals of Science* 35 (1978): 34–44.

Levy, Paul. *Moore: G. E. Moore and the Cambridge Apostles*. London: Weidenfeld and London, 1979.

Lindemann, Ferdinand. "Über die Zahl." *Mathematische Annalen* 20 (1882): 213–225.

Lobachevsky, Nikolai. *Geometrical Researches on the Theory of Parallels*. Translated by George Bruce Halsted. Austin: The University of Texas Press, 1891.

Locke, John. *An Essay Concerning Human Understanding*. Oxford: Clarendon Press, 1975.

London Mathematical Society. *List of Members of the London Mathematical Society, 13th November, 1884*. London: C. F. Hodgson & Son, 1884.

Losee, J. "Whewell and Mill on the Relations between Philosophy of Science and History of Science." *Studies in the History and Philosophy of Science* 14 (1979): 113–126.

Lowe, Victor. *Alfred North Whitehead: The Man and His Work*. Baltimore: Johns Hopkins University Press, 1985.

Macfarlane, Alexander. *Lectures on Ten British Mathematicians of the Nineteenth Century*. New York: Wiley, 1916.

MacHale, Desmond. *George Boole: His Life and Work*. Dublin: Boole Press, 1985.

Machin, G. I. T. *Politics and the Churches in Great Britain 1832–1868*. Oxford: Clarendon Press, 1977.

Malebranche, Nicolas de. *The Search After Truth*. Translated and edited by Thomas M. Lennon and Paul J. Olscamp. Cambridge: Cambridge University Press, 1997.

McFarland, Thomas. *Coleridge and the Pantheist Tradition*. Oxford: Clarendon Press, 1969.

Mehrtens, Herbert, Henk Bos, and Ivo Schneider, eds. *Social History of Nineteenth Century Mathematics*. Boston: Birkhäuser, 1981.

Merlan, Philip. *From Platonism to Neoplatonism*. The Hague: Martinus Nijhoff, 1975.

Merrill, Daniel D. *Augustus De Morgan and the Logic of Relations*. Dordrecht: Kluwer Academic Publishers, 1990.

Mill, John Stuart. *Autobiography*. London: Longmans, Green, Reader, and Dyer, 1873.

———. *An Examination of Sir William Hamilton's Philosophy and of the Principle*

Philosophical Questions Discussed in his Writings, 2nd ed. London: Longmans, Green, and Co., 1865.

———. *A System of Logic Ratiocinative and Inductive: Being a Connected View of the Principles of Evidence and the Methods of Scientific Investigation*. New York: Harper & Brothers, 1846.

Moore, Edward D. O. *Geometrical Science*. New York: The Moss Engraving Co., 1890.

Morrell, Charles. *The Squaration of the Circle*. Chicago: The Hesperan VE, 1924.

Morrell, Jack, and Arnold Thackray. *Gentlemen of Science: Early Years of the British Association for the Advancement of Science*. Oxford: Clarendon Press, 1981.

Morrell, Jack, and Arnold Thackray, eds. *Gentlemen of Science: Early Correspondence of the British Association for the Advancement of Science*. London: Royal Historical Society, 1984.

Muirhead, John H. *The Platonic Tradition in Anglo-Saxon Philosophy*. London: George Allen & Unwin, 1931.

Myers, William Alexander. *The Quadrature of the Circle, the Square Root of Two, and the Right-Angled Triangle*. Cincinnati: Wilstach, Baldwin & Co., 1873.

Nagel, Ernst. "'Impossible Numbers': A Chapter in the History of Modern Logic." In *Studies in the History of Ideas*, edited by the Columbia University Philosophy Department, 429–474. New York: Columbia University Press, 1935.

New York Mathematical Society. *New York Mathematical Society List of Members, Constitution, By-Laws*. New York: Published by the Society, 1892.

Newman, Francis W. *Phases of Faith*. London: John Chapman, 1850.

Newman, John Henry. *The Idea of a University*, edited by Frank M. Turner. New Haven, CT: Yale University Press, 1996.

Nichol, J. P. *The Planet Neptune*. Edinburgh: James Nichol, 1849.

Nichols, Ichabod. *A Catechism of Natural Theology*. Boston: William Hyde, 1831.

———. *An Oration*. Boston: Joshua Cushing, 1805.

Norris, John. *An Essay Towards the Theory of the Ideal or Intelligible World*. London: S. Manship and W. Hawes, 1701–1704.

Olsen, Alexandra, and John Voss, eds. *The Organization of Knowledge in Modern America 1860–1920*. Baltimore: Johns Hopkins University Press, 1979.

Olson, Richard. *Scottish Philosophy and British Physics, 1750–1880: A Study in the Foundations of the Victorian Scientific Style*. Princeton, NJ: Princeton University Press, 1975.

———. "Scottish Philosophy and Mathematics: 1750–1830." *Journal of the History of Ideas* 32 (1971): 29–44.

Oppenheim, Janet. *The Other World: Spiritualism and Psychical Research in England, 1850–1914*. Cambridge: Cambridge University Press, 1985.

Orsini, G. N. *Coleridge and German Idealism*. Carbondale, IL: Southern Illinois University Press, 1969.

Parker, John A. *Quadrature of the Circle*, rev. and enl. ed. New York: John Wiley & Son, 1874.

Parker, Theodore. *The Transient and Permanent in Christianity.* Boston: Beacon Press, 1948.

Parshall, Karen Hunger. "America's First School of Mathematical Research: James Joseph Sylvester at The Johns Hopkins University 1876–1883." *Archive for History of Exact Sciences* 38 (1988): 153–196.

Parshall, Karen Hunger, and David E. Rowe, *The Emergence of the American Mathematical Research Community, 1876–1900: J. J. Sylvester, Felix Klein, and E. H. Moore.* Providence, RI: The American Mathematical Society and the London Mathematical Society, 1994.

Parsons, Gerald, ed. *Religion in Victorian Britain.* Manchester: Manchester University Press, 1988.

Pascal, Blaise. *Pensées.* Translated by A. J. Krailsheimer. Harmondsworth, U.K.: Penguin Books, 1966.

Pattison, F. L. M. *Granville Sharp Pattison.* Tuscaloosa, AL: University of Alabama Press, 1987.

Peabody, Andrew P. *Elegy at the Funeral of Benjamin Peirce.* Boston, 1880.

———. *Harvard Reminiscences.* Boston: Ticknor and Co., 1888.

———. *Sermon Preached at Portland, Maine, at the Funeral of Rev. Ichabod Nichols, D.D.* Boston: Crosby, Nichols, 1859.

Peacock, George. *Observations on the Statutes of the University of Cambridge.* London: J. W. Parker, 1841.

Pearson, Karl. "Old Tripos Days at Cambridge, as Seen from Another Viewpoint." *The Mathematical Gazette* 20 (1936): 27–36.

Peirce, Benjamin. *Address to the American Association for the Advancement of Science.* Washington, DC: Printed by the Association, 1853.

———. "The Conflict Between Science and Religion." *Unitarian Review* 7 (June 1877): 655–666.

———. *An Elementary Treatise on Algebra: To Which Are Added Exponential Equations and Logarithms.* Boston: J. Munroe & Co., 1837.

———. *An Elementary Treatise on Plane and Solid Geometry.* Boston: J. Munroe & Co., 1837.

———. *Ideality in the Physical Sciences.* Boston: Little, Brown, and Co., 1881.

———. "The Intellectual Organization of Harvard." *Harvard Register* 1 (1880): 77.

———. *Linear Associative Algebra.* New York: D. Van Nostrand, 1870; reprint, New York: D. Van Nostrand, 1882.

Peirce, Charles Sanders. *Collected Papers.* Cambridge, MA: Harvard University Press, 1934.

Percival, Janet, ed. *The Society for the Diffusion of Useful Knowledge, 1826–1848.* London: University College London Library, 1978.

Perry, John. "The Mathematical Tripos at Cambridge." *Nature* 75 (1907): 273–274.

Plato. *Opera.* Translated by Thomas Taylor. London: R. Wilks, 1804.

———. *Philebus and Epinomis.* Translated by A. E. Taylor. London: Thomas Nelson and Sons, 1956.

———. *The Republic.* Translated by G. M. A. Grube. Indianapolis, IN: Hackett, 1974.

———. *Timaeus.* Translated by Francis M. Cornford. New York: Macmillan, 1959.

Prantl, Karl. *Geschichte der Logik im Abendlande.* Leipzig, 1855–1870.

Prickett, Stephen. *Romanticism and Religion: The Tradition of Coleridge and Wordsworth in the Victorian Church.* Cambridge: Cambridge University Press, 1976.

Proclus. *The Philosophical and Mathematical Commentaries of Proclus on the First Book of Euclid's Elements.* Translated by Thomas Taylor. London, 1789.

Pycior, Helena. "At the Intersection of Mathematics and Humor: Lewis Carroll's *Alices* and Symbolical Algebra." *Victorian Studies* 28 (1984): 149–170.

———. "Augustus De Morgan's Algebraic Work: The Three Stages." *Isis* 74 (1983): 211–226.

———. "Early Criticism of the Symbolical Approach to Algebra." *Historia Mathematica* 9 (1982): 393–412.

———. "George Peacock and the British Origins of Symbolical Algebra." *Historia Mathematica* 8 (1981): 23–45.

———. "Internalism, Externalism, and Beyond: 19th Century British Algebra." *Historia Mathematica* 11 (1984): 424–441.

Ralls, Walter. "The Papal Aggression of 1850: A Study in Victorian Anti-Catholicism." In *Religion in Victorian Britain,* edited by Gerald Parsons, 115–134. Manchester: Manchester University Press, 1988.

Rantoul, Robert S. *Memoir of Benjamin Peirce.* Salem, MA: Essex Institute, 1881.

Reardon, Bernard M. G. *From Coleridge to Gore: A Century of Religious Thought in Britain.* London: Longmans, 1971.

Recordon, C. J. *Solutions Approchées de la Trisection de L'Angle et de la Quadrature du Cercle.* Paris: Gauthier-Villars, 1865.

Reingold, Nathan, ed. *Science in Nineteenth-Century America.* Chicago: University of Chicago Press, 1985.

Rescher, Nicholas. *Pascal's Wager.* Notre Dame, IN: University of Notre Dame Press, 1985.

Richards, Joan L. *Angles of Reflection.* New York: W. H. Freeman, 2000.

———. "The Art and the Science of British Algebra: A Study in the Perception of Mathematical Truth." *Historia Mathematica* 7 (1980): 343–365.

———. "Augustus De Morgan, the History of Mathematics, and the Foundations of Algebra." *Isis* 78 (1987): 6–30.

———. "God, Truth, and Mathematics in Nineteenth Century England." In *The Invention of Physical Science,* edited by Mary Jo Nye et al., 51–78. Dordrecht: Kluwer Academic Publishers, 1992.

———. *Mathematical Visions: The Pursuit of Geometry in Victorian England.* Boston: Academic Press, 1988.

———. "Projective Geometry and Mathematical Progress in Mid-Victorian Britain." *Studies in the History and Philosophy of Science* 17 (1968): 297–325.

———. "The Reception of a Mathematical Theory: Non-Euclidean Geometry in England, 1868–1883." In *Natural Order: Historical Studies of Scientific Culture,* edited by Barry Barnes and Stephen Shapin, 143–166. Beverly Hills, CA: Sage Publications, 1979.

Richards, John. "Boole and Mill: Differing Perspectives on Logical Psychologism." *History and Philosophy of Logic* 1 (1980): 19–36.

Richardson, R. G. D. "The Ph.D. Degree and Mathematical Research." *American Mathematical Monthly* 43 (1936): 199–215.

Richardson, Robert D. *Emerson: The Mind on Fire.* Berkeley: University of California Press, 1995.

Richey, Russell E. *Origins of English Unitarianism.* Princeton, NJ: Princeton University Press, 1970.

Riemann, Bernhard. "On the Hypotheses which Lie at the Bases of Geometry." Translated by William Kingdon Clifford. *Nature* 8 (1873): 14–17, 36–37.

Roberts, Jon. *Darwinism and the Divine in America.* Madison: University of Wisconsin Press, 1988.

Rosenfeld, Boris A. *A History of Non-Euclidean Geometry.* Translated by Abe Shenitzer. New York: Springer-Verlag, 1988.

Rothblatt, Sheldon. *The Revolution of the Dons: Cambridge and Society in Victorian England.* New York: Basic Books, 1968.

———. *Tradition and Change in English Liberal Education.* London: Faber and Faber, 1976.

Rowe, David E., and John McCleary, eds. *The History of Modern Mathematics.* Boston: Academic Press, 1989.

Rowse, A. L. *The Controversial Colensos.* Trewolsta, U.K.: Cornish Publications, 1989.

Russell, Bertrand. *The Autobiography of Bertrand Russell, 1872–1914.* Boston: Little, Brown, and Co., 1967.

———. *The Collected Papers of Bertrand Russell.* London: George Allen & Unwin, 1983.

———. *An Essay on the Foundations of Geometry.* New York: Dover, 1956.

———. *A History of Western Philosophy.* New York: Simon and Schuster, 1945.

———. *My Philosophical Development.* London: George Allen and Unwin, 1959.

———. *Mysticism and Logic.* New York: Longmans, Green, 1918.

———. *The Principles of Mathematics,* 2nd ed. New York: W.W. Norton & Co., 1938.

———. "Recent Work on the Principles of Mathematics." *International Monthly* 4 (1901): 83–101.

———. "A Turning Point in My Life." *The Saturday Book* 8 (1948): 142–146.

Rutherford, William. "Computation of the Ratio of the Diameter of a Circle to its Circumference to 208 Places of Figures." *Philosophical Transactions of the Royal Society of London* 131 (1841): 281–283.

———. "On the Extension of the Value of the Ratio of the Circumference of a Circle to its Diameter." *Proceedings of the Royal Society of London* 6 (1853): 273–275.

Sarton, George. *The Study of the History of Mathematics.* Cambridge, MA: Harvard University Press, 1937.

Schepler, Herman C. "The Chronology of Pi." *Mathematics Magazine* 23 (March/ April and May/June 1950): 216–228, 279–283.

Schrickx, W. "Coleridge and the Cambridge Platonists." *Review of English Literature* 7 (1966): 71–91.

Scott, J. F. *The Mathematical Work of John Wallis.* London: Taylor and Francis, 1938.

Seaton, Ethel. "Thomas Harriot's Secret Script." *Ambix* 4 (October 1956): 111–114.

Shanks, William. *Rectification of the Circle.* London: G. Bell, 1853.

Shearman, A. T. *The Development of Symbolic Logic.* London: Williams and Norgate, 1906.

Shirley, John W. *Thomas Harriot: A Biography.* Oxford: Clarendon Press, 1983.

Smith, G. C., ed. *The Boole-De Morgan Correspondence 1842–1864.* Oxford: Clarendon Press, 1982.

———. "De Morgan and the Laws of Algebra." *Centaurus* 25 (1981): 50–70.

Smith, H. J. Steven. "On the Present State and Prospects of Some Branches of Pure Mathematics." *Proceedings of the London Mathematical Society* 8 (1876–1877): 6–29.

Smith, H. Shelton. "Was Theodore Parker a Transcendentalist?" *New England Quarterly* 23 (1950): 1–32.

Smith, James. *A Nut to Crack for the Readers of Prof. De Morgan's 'Budget of Paradoxes.'* London: Simpkin & Marshall, 1863.

———. *The Problem of Squaring the Circle Solved.* London: Longman, Brown, Green, Longmans, & Roberts, 1859.

———. *Quadrature du Cercle.* Translated by Armand Granges. Bordeaux: Lafargue, 1863.

———. *The Quadrature of the Circle.* London: Simpkin, Marshall & Co., 1865.

———. *The Quadrature of the Circle: Correspondence between an Eminent Mathematician and James Smith, Esq.* London: Simpkin, Marshall, & Co., 1861.

Snyder, Alice D. *Coleridge on Logic and Learning.* New Haven, CT: Yale University Press, 1929.

Somerville, Mary. *Personal Recollections.* Boston: Roberts Brothers, 1876.

Souleymane, Bachir Diagne. *Boole, 1815–1864: L'Oiseau de Nuit en Plein Jour.* Paris: Berlin, 1989.

Standage, T. *The Neptune File: A Story of Astronomical Rivalry and the Pioneers of Planet Hunting.* New York: Walker and Co., 2000.

Stewart, Balfour, and Peter Guthrie Tait. *The Unseen Universe; or Physical Speculations on a Future State.* London: Macmillan, 1875.

Stirling, A. M. W. *William De Morgan and His Wife.* New York: Henry Holt and Co., 1922.

Stokes, George Gabriel. *Mathematical and Physical Papers.* Cambridge: Cambridge University Press, 1905.

———. *Memoir and Scientific Correspondence of the Late Sir George Gabriel Stokes,* edited by Joseph Larmor. Cambridge: Cambridge University Press, 1907.

———. *Natural Theology*. London: Adam and Charles Black, 1893.

Storr, Richard J. *The Beginnings of Graduate Education in America*. Chicago: University of Chicago Press, 1953.

Story, William E. "On the Non-Euclidean Geometry." *American Journal of Mathematics* 5 (1882): 180–211.

Styazhkin, N. I. *History of Mathematical Logic from Leibniz to Peano*. Cambridge, MA: MIT Press, 1969.

Sylvester, James Joseph. *The Collected Mathematical Papers of James Joseph Sylvester*. Cambridge: Cambridge University Press, 1904–1912.

———. "On the Theory of the Syzygetic Relations of Two Rational Integral Functions, Comprising an Application to the Theory of Sturms Functions, and that of the Greatest Algebraical Common Measure," *Philosophical Transactions of the Royal Society of London* 143 (1853): 407–548.

———. "Presidential Address: Mathematics and Physics Section." In *Report of the Thirty-Ninth Meeting of the British Association for the Advancement of Science held at Exeter in August, 1869*, 1–9. London: John Murray, 1870.

Taylor, Geoffrey. "George Boole, F.R.S., 1815–1864." *Notes and Records of the Royal Society of London* 12 (1956): 44–52.

Taylor, Thomas. *Thomas Taylor the Platonist: Selected Writings*, edited by Kathleen Raine and George Mills Harper. Princeton, NJ: Princeton University Press, 1969.

Thomas, Donald. *Lewis Carroll*. London: John Murray, 1996.

Thomas, John Meurig. *Michael Faraday and the Royal Institution*. Bristol, U.K.: Adam Hilger, 1991.

Todhunter, Isaac. *The Conflict of Studies and Other Essays on Subjects Connected with Education*. London: Macmillan, 1873.

———. *William Whewell: An Account of His Writings, with Selections from his Literary and Scientific Correspondence*. 1876; reprint, New York: Johnson Reprint Corp., 1970.

Torretti, Roberto. *Philosophy of Geometry from Riemann to Poincaré*. Dordrecht: D. Reidel, 1978.

Townsend, G. H., ed. *Men of the Time*. London: George Routledge and Sons, 1868.

Turner, Frank M. *Between Science and Religion: The Reaction to Scientific Naturalism in Late Victorian England*. New Haven, CT: Yale University Press, 1974.

———. "The Victorian Conflict between Science and Religion: A Professional Dimension." *Isis* 69 (1978): 356–376.

Tyndall, John. *Faraday as a Discoverer*. London: Longmans, Green, 1894.

Van Der Weyde, P. H. *The Quadrature of the Circle Demonstrated to be Perfectly Soluble by Modern Mathematics*. New York: D. Appleton, 1861.

Van Evra, James. "Richard Whately and the Rise of Modern Logic." *History and Philosophy of Logic* 5 (1984): 1–18.

Veitch, John. *Memoir of Sir William Hamilton, Bart*. Edinburgh and London: William Blackwood and Sons, 1869.

Venn, John. *The Logic of Chance*. London: Macmillan, 1866.

————. *Symbolic Logic*. London: Macmillan, 1881.

Vericour, Louis Raymond de. *An Historical Analysis of Christian Civilisation*. London: John Chapman, 1850.

Veysey, Laurence R. *The Emergence of the American University*. Chicago: University of Chicago Press, 1965.

Wait, Lucien A. "Advanced Instruction in American Colleges." *The Harvard Register* 3 (1881): 127.

Wallace, Richard. *The Agony of Lewis Carroll*. Melrose, MA: Gemini Press, 1990.

Wallis, John. *Treatise of Algebra*. London, 1685.

Wallis, R. T. *Neoplatonism*. London: Duckworth, 1972.

Ware, Henry, Jr. *The Works of Henry Ware Jr., D.D.* Boston: James Monroe and Company, 1846.

Weaver, Warren. "Lewis Carroll: Mathematician." *Scientific American* 194 (April 1956): 116–128.

Webb, R. K. "The Limits of Religious Liberty: Theology and Criticism in Nineteenth-Century England." In *Freedom and Religion in the Nineteenth Century*, edited by Richard Helmstadter, 120–149. Stanford: Stanford University Press, 1997.

Weiss, John. *Discourse Occasioned by the Death of Convers Francis, D.D.* Cambridge, 1863.

Welch, Claude. *Protestant Thought in the Nineteenth Century*. New Haven, CT: Yale University Press, 1972–1985.

Whately, Richard. *Elements of Logic*. New York: William Jackson / Boston: James Munroe & Co., 1836.

Whewell, William. *Astronomy and General Physics Considered with Reference to Natural Theology*. London: W. Pickering, 1833.

————. *The Doctrine of Limits, with its Applications*. Cambridge: J. and J. J. Deighton, 1838.

————. *History of the Inductive Sciences from the Earliest to the Present Time*, 3rd ed. London: John W. Parker, 1837.

————. *Indications of the Creator: Extracts Bearing upon Theology, from the History and Philosophy of the Inductive Sciences*. London: John W. Parker, 1845.

————. *Influence of the History of Science upon Intellectual Education*. Boston: Gould and Lincoln, 1854.

————. *Of a Liberal Education in General; and with Particular Reference to the Leading Studies of the University of Cambridge*. London: John W. Parker, 1845.

————. *The Philosophy of the Inductive Sciences Founded upon Their History*, 2nd ed. London: John W. Parker, 1847.

————. *Thoughts on the Study of Mathematics as a Part of a Liberal Education*. Cambridge, 1835.

Whitehead, Alfred North. *Science and the Modern World*. New York: Macmillan, 1925.

————. *A Treatise on Universal Algebra*. Cambridge: Cambridge University Press, 1898.

Whitehead, Alfred North, and Bertrand Russell. *Principia Mathematica,* 2nd ed. Cambridge: Cambridge University Press, 1925.

Wilkins, John. *Mathematical and Philosophical Works.* London, 1802.

Willerding, Margaret F. "Mathematics Through a Looking Glass." *Scripta Mathematica* 25 (November 1960): 209–219.

Williams, Leslie Pearce. *Michael Faraday, A Biography.* New York: Basic Books, 1965.

Willis, William. "Inaugural Address of the President to the Maine Historical Society, March 1857," in *Collections of the Maine Historical Society.* Portland: Maine Historical Society, 1857.

Willmon, Jeremy C. *The Secret of the Circle and the Square.* Los Angeles: McBride Press, 1905.

Willson, F. M. G. *Our Minerva: The Men and Politics of the University of London, 1836–1858.* London: Athlone, 1995.

Wolterstorff, Nicholas. *John Locke and the Ethics of Belief.* Cambridge: Cambridge University Press, 1996.

Wordsworth, William. *The Prelude, or Growth of a Poet's Mind.* London: Oxford University Press, 1805; reprint, London: Oxford University Press, 1933.

Wright, Conrad. *The Beginnings of Unitarianism in America.* Boston: Starr King Press, 1955.

———. *The Liberal Christians: Essays on American Unitarian History.* Boston: Beacon Press, 1970.

Yates, Francis. *The Art of Memory.* London: Routledge & K. Paul, 1966.

Yeo, Richard. *Defining Science: William Whewell, Natural Knowledge, and Public Debate in Early Victorian Britain.* Cambridge: Cambridge University Press, 1993.

Index

202n64; and professionalization of mathematics, 12–13, 106–7, 136, 141–42, 145, 160–61, 174; religious views of, 40, 106, 109–10, 113–14, 202n67; and spiritualism, 109, 125–31; and symbolic logic, 10, 11, 109, 121–25, 134, 204n88; and University College, London, 110, 111, 112, 115–17, 157, 158–59
De Morgan, George, 106
De Morgan, Sophia Frend, 110, 113, 115, 126–27
Descartes, René, 133, 139
Diderot, Denis, 161
distribution, law of, 100
Dodgson, Charles Lutwidge, 172–76, 181, 212–13n140

ecumenism: of Augustus De Morgan, 5, 11, 12, 27, 40–41, 106, 109, 127; of George Boole, 11, 27, 38–41, 80, 92–94, 111; and mathematical idealism, 52, 76, 93, 94; and symbolic logic, 11, 27–29, 98. *See also* antidogmatism
Edinburgh Mathematical Society, 146
educational reform, 62–67
Einstein, Albert, 164
Elements (Euclid), 21, 106, 138, 179, 180, 214n185; John Dee's introduction to, 21–23. *See also* Euclid
Eliot, Charles, 65, 66
Emerson, Ralph Waldo, 45, 47–50, 59, 66, 190n37; and Benjamin Peirce, 42, 45, 47–48, 75
Epinomis (Plato), 18–19, 62, 73
Euclid, 165; *Elements* by, 21, 106, 138, 179, 180, 214n185; long-lasting influence of, 21, 58, 79, 147, 164, 174; in mathematics education, 147, 148, 214n185; waning influence of, in late Victorian era, 138, 148, 164, 178–79, 180 (*see also* non-Euclidian geometry)
Euclid and his Modern Rivals (Dodgson), 173–74
Euler, Leonhard, 49, 161

Everest, George, 79
Everest, Mary. *See* Boole, Mary Everest
Everett, Edward, 8, 67
evolution, 68–69, 70, 138

Faraday, Michael, 109, 128–29, 131
Farrar, Adam S., 8
Ferrers, N. M., 148
Ficino, Marsilio, 21
Finkel, B. F., 150
Formal Logic (De Morgan), 10, 121, 122–23, 133, 156
Fortnightly Review, The, 166
Francis, Convers, 46
Frege, Gottlob, 179
French Academy, 3–4, 139
French Revolution, 52
Frend, William, 115
Friedrich, Caspar David, 37–38
Froude, James Anthony, 113
Fuller, Rufus, 153

Galileo, 18, 55, 185n20
Galle, Johann Gottfried, 1
Gentleman's Diary, The (periodical), 147, 175
geology, 53
geometry, 71, 92, 138, 147; and discovery of Neptune, 54, 92; exalted claims for, 54, 63, 68, 72–73; teaching of, 63–64, 147
German philosophical idealism, 45–46; influence of, in Britain and United States, 8–9, 29–31, 45–46, 48, 120. *See also* Kant, Immanuel
Germany, 176, 179. *See also* German philosophical idealism
Gifford, Lord Adam, 168
Glas, John, 128
Goldsmid, Isaac Lyon, 110
Göttingen University, 176
Greeks, ancient, 5, 8, 16, 55–56, 72–73. *See also* Archimedes; Aristotle; Euclid; Plato
Gregg, Tresham, 155–56, 157, 160